Sustainable Solid Waste Management

The Author

URVASHI DHAMIJA grew up in Delhi. She is a Reader, Department of Political Science, Miranda House, University of Delhi. She is a founder member of *M.H. Vatavaran*, the society for environmental concerns, in the college.

sustainable solid waste management

ISSUES POLICIES AND STRUCTURES

Urvashi Dhamija

Academic Foundation
NEW DELHI

Published in 2006 by

ACADEMIC FOUNDATION

4772-73 / 23 Bharat Ram Road, (23 Ansari Road),
Darya Ganj, New Delhi - 110 002 (India).
Phones : 23245001 / 02 / 03 / 04.
Fax : +91-11-23245005.
E-mail : academic@vsnl.com
www.academicfoundation.com

Sustainable Solid Waste Management: Issues, Policies and Structures
by Urvashi Dhamija

ISBN 81-7188-428-8

Designed and typeset by Italics India, New Delhi.
Printed and bound in India

This book is dedicated to Miranda House where ideas
"whose time has come" find expression

Contents

...CONTD. ...

List of Figures and Tables

FIGURES

TABLES

Acknowledgements

This work was made possible because the Governing Body of Miranda House granted me leave to study for it during 2002. I am grateful to Professor Mohan Ram the then Chairman of the Governing Body, Dr. Surjeet Suneja the Acting Principal of Miranda House and to Professor Neera Chandhoke, Head of the Department of Political Science, University of Delhi who recommended the grant of the leave.

I am indebted to my former colleagues, Professor Anu Kapur, Department of Geography, Delhi School of Economics; Professor Niraja Gopal Jayal, Centre of Law and Governance and to Dr. Shefali Jha, Assistant Professor, Centre of Political Studies, Jawaharlal Nehru University who helped me clarify my ideas on how a coherent statement could be attempted on an issue of persistent private anxiety. Niraja also very kindly agreed to write the Foreword for this book.

I am grateful to Professor Manoranjan Mohanty, Department of Political Science, University of Delhi and to Professor George Koreth, Director, Asian Centre for Organisation Research and Development (New Delhi), who generously gave me their time when this work was in progress. They read through sections of the draft and made extremely helplful observations.

I want to thank Dr. Shipra Maitra of the Institute of Human Development, Dr. Rakesh Gupta of the Institute of Public Administration, Mr. Ravi Agarwal and his colleagues at Toxic Links, Dr. Shyamala Mani of the Centre for Environment Education and Dr Suneel Pandey of the Tata Energy Research Institute for sharing their insights about the difficulties in Solid Waste Management based on their study, research and practical involvement in this field.

I profited immensely from my discussions with Mr. Uttam Vasvani, Joint Director Conservancy and Sanitation Engineering (CSE) Division, Municipal Corporation of Delhi (MCD), Mr. S.K. Mahajan, Officer on Special Duty, Yamuna Action Plan, Mr. Ravi Dass, Director,

Mr. P.K. Khandelwal and Mr. Surinder Pal, Executive Engineers in the CSE Division of the MCD and with Dr Anil Kumar and Dr Sabata of the Department of Environment, Government of the National Capital Territory of Delhi.

I want to record my appreciation of the assistance I got in collecting data from Mr. Ainul Abedin and Ms. Qamar Raza, members of the Library Staff of the Habitat Centre, New Delhi and from Mrs Uma Ramachandra, Information Resource Coordinator, World Bank, New Delhi.

My father, Mr. R.N. Vasudeva and my mother Mrs. Santosh Vasudeva, my husband Vinod, and my children Isha and Shantanu gave me continuous encouragement and support. Without their help this project would not have been completed.

Finally, I want to express my gratitude towards staff of the Academic Foundation and particularly to the members of the Kapila family. It is their counsel, assistance and patience which has enabled an elusive goal to become a tangible reality.

Urvashi Dhamija

Foreword

I am delighted to write this foreword to Urvashi Dhamija's important new book *Sustainable Solid Waste Management: Issues, Policies and Structures*. This is a wide-ranging and comprehensive study of the issue of solid waste management in the city of Delhi between 1998 and 2003. The capital city of India, Delhi has many distinctive attributes: a high population density, to which high rates of in-migration contribute half a million people annually; many different types of settlements, from slums and unauthorised colonies to planned colonies; a multiplicity of agencies of governance, resulting in competing jurisdictions; and, as a result of governmental initiative, the impressive participation of non-governmental organisations. Dhamija situates her study of solid waste management in this complex and varied context.

Taking as its benchmark the post-Rio practices of sustainable solid waste management, the study examines four types of waste: non-hazardous waste, and hazardous waste of three kinds, *viz.* toxic chemical waste, bio-medical waste, and human waste. The kind of waste generated in a city, this book shows us, signifies the type of city it is, and the break-up of Delhi's waste tells us that a large proportion of its residents are poor. Dhamija's research shows how the efforts of the Government of Delhi between 1998-2003—in the backdrop of judicial intervention, and complemented by policies of the Union government—resulted in the successful adoption of practices of sustainable solid waste management. Though much still remains to be accomplished, it is important to take note of the successes achieved, and of what made them possible. Dhamija shows us how the Delhi Government's desire to make the city more "clean and green" resulted in enlightened policy initiatives, such as the promotion of non-burn technology for the management of bio-medical waste. The most important component of this success, however, was an institutional innovation in governance: the participation of non-governmental organisations, and of citizen groups in the Bhagidari initiative of the Delhi Government.

Based upon this analysis, the author makes policy recommendations for achieving even higher standards of sustainable solid waste management. These include suggestions for institutional reform, and the importance of enhancing the transparency and accountability of governance through mechanisms such as the Right to Information.

Though her focus is on only one aspect of environmental management, Dhamija has successfully illuminated a variety of issues. This book enhances our understanding of both the relatively technical issues of environmental degradation, as well as the issues of policy, institutional reform and governance, and the links between these. It should attract a wide readership of academics, policy makers, members of civil society organisations and not least the civically minded citizen.

Niraja Gopal Jayal

Professor, Centre for the Study of Law and Governance,
Jawaharlal Nehru University (JNU),
New Delhi.

Introduction

Contemporary Delhi, India's capital city, long standing hub of politics and administration is situated in the same general area from where rulers have functioned off and on since 12th century A.D. It is now a multi-functional city with large-scale industrial, commercial and service-oriented activities. With 1000 people being added to the population every day, it is one of India's fastest growing cities. Its huge size—13.7 million live here—makes it part of a select group of cities, the 'mega' cities of the developing world that have 10 million and above residents. The growth rate of Delhi's population at 3.8 per cent per annum, is twice that of the national average. In this sense, Delhi is similar to Dhaka which grew in population by an average of nearly 7 per cent a year from 1975–2000 compared to an annual average of 2.1 per cent for Bangladesh as a whole and also to Lagos which grew at an average of 5.6 per cent per year compared to 3 per cent for Nigeria as a whole. Delhi's growth at the expense of other urban centres has been due to the range of opportunities for economic growth that it offers to citizens in other parts of the country and especially to the poor in the neighbouring states of UP and Bihar due to the growing demand for manpower in the tertiary sector. The per capita income here, is more than double the national average. As compared to 29 other States and 6 Union Territories, more people in Delhi possess a TV set, use LPG and are likely to use their home as a shop or as an office. Delhi is possibly India's most pragmatism driven multi-class and multi-ethnic city.

Some years ago, Delhi was given an international rank for an entirely unenviable reason—it was considered the world's third most polluted city. Over the years, apart from improvement in its air quality, there has been some reduction in the filth and squalor on its streets. Since January 2004, there has been a rapid growth in the number of specially designed brightly coloured receptacles for collecting waste in public places. But from a capital city of a continental sized country more is expected. By the logic of its premier position, it is a showcase of what can be achieved not only by urban centres within the country but also those abroad. As a dynamic city it is a potential source of indicators of how a developing country can adapt to the challenges of 'modernisation' and 'liberalisation.'

Solid waste is unwanted material, a by-product of production and consumption abandoned within municipal limits for whose safe disposal, government agencies are responsible. The task of managing waste better in the case of Delhi is crucial for at least three reasons. Its unique position deserves to be leveraged to demonstrate the close link between cleaner surroundings, greater productivity and higher tourist inflow. Secondly, since waste removal is a low level 'polluting' function, political prioritisation will help in countering narrow caste based attitudes and strengthening secular egalitarian trends in keeping with the values of India's Constitution. Finally, in an era of relentless global pressures, changes in waste management arrangements offer an opportunity to demonstrate how deference towards local knowledge, initiative and livelihood concerns can be combined with commitment towards state-of-art technology, and new forms of international partnership. 'Glo-calism' can be demonstrated as a viable guiding principle for the future.

In Delhi, 6,000-7,000 tonnes of waste is generated every day and it is expected that on the basis of growth assumption of 6 per cent to 8 per cent (economic growth plus 2 per cent), the quantity of waste will be 17,000 to 25,000 tonnes a day in 2021. The persistent inadequacies of the municipal body, the legally authorised entity in carrying out the collection, transportation and safe disposal of waste together with increasing evidence of hazards to health and the environment, prevention of which is very purpose of solid waste management, suggests that the time has come for a comprehensive review of the issues, policies and structures involved in these operations.

The substantive issues in making cities in developing countries garbage free in a durable way are more complex than what they seem on the surface. This is because the composition of waste is very heterogeneous, and the management of the logistics of its different components poses enormous practical problems. City waste for policy-making purposes is regarded as being non-hazardous. This is because it consists largely of organic waste and non-bio-degradable consumer waste from homes and offices. Potentially hazardous waste, such as human excreta and toxic waste, such as soiled bandages from health care centres and danger laden chemical residues from factories that are co-disposed with municipal waste are ignored because they comprise a tiny fraction of the total waste. With growth of industrialisation and urbanisation there has been an increase in the generation of all categories of waste and a huge lag in the creation of appropriate facilities required for their safe disposal. All types of waste is simply dumped in municipal areas whether these are streets, drains or sewers. Improvements in the management of solid waste has not had much visibility or experiential impact despite the huge expenditure that is regularly made in carrying these out because official policy has focused largely on the final disposal of the increasing levels of food scrap and packaging materials in municipal waste. Debates and innovations have revolved around segregation of household and institutional waste into bio-degradable and non-bio-degradable waste, modification of designs of public receptacles to exclude cattle and rag-pickers, replacement of human being by machines for loading waste and increase of frequency of trips for secondary collection etc. Adverse social and environmental implications of human and hazardous waste in municipal solid waste get neglected because neither policy nor infrastructure is geared to deal with it. In Bichchri, Rajasthan, the estimated cost of undoing damage by chemicals to 350 acre was Rs. 44 crore. Against this background it can be said that the substantive issue in durable or sustainable waste management is ensuring that all government agencies that impact on a city's affairs, central regional, sub-regional and local act in tandem so that all types of waste, non-hazardous and hazardous, from homes, institutions, health centres and factories are handled and routinely disposed of in a manner which is safe both from the point of view of human-beings and the surrounding environment.

The task of adapting governmental policies is not easy despite data repeatedly showing a gap between objectives and achievements because existing policies involve entrenched legal, structural, technical, social and financial aspects. In Indian cities this is most clearly evident in relation to human waste. The basic system supported by the Central government and the State governments for its safe disposal consists of extensions into residential areas of sewer pipes that transport excreta laden water into distant high-tech wastewater treatment plants where the solids are removed and the water in some state of cleanliness is released into the nearest lake, river or sea. In 2000, only 30 per cent of the urban areas in India—55 per cent in the case of Delhi—were covered by this 'centralised' or alternatively, a 'settled' sewerage system. In the second case in anticipation of construction of a sewer line in the future, waste is collected in a 'soak pit' or septic tank from where it is manually or mechanically removed. Even though this is a capital intensive system which requires increasingly scarce drinking quality water for its maintenance and can be regarded as the main cause of death among children under five years and sickness among adults because its limited reach means that large numbers are deprived of adequate sanitation, decision-makers have found it impractical to move away from this model. There is only token support for less material and capital intensive options because no alternative system can match the present one in creating a hygienic human waste disposal system on a similar scale. Yet, if epidemics are to be avoided and lakes and rivers, the sources of fresh water are to be prevented from becoming sewers, then present policies will have to be modified to fund an aggressive research effort for inventing a low cost, ecologically benign, culturally sensitive toilet which is appropriate for the crowded conditions of urban living.

The primary source of influence of a governmental structure is the constitutional-legal framework within which it is located. It is generally said that since the problem of waste has become so huge and complex, governmental agencies should 'involve the people' and from 'a providers of solutions,' become a 'facilitators.' The fact is that even when there are changes in the law to re-define the authority of a governmental office and equip it for exercising its responsibilities in an environment where there are more players, for example, by empowering it to sign contracts with private companies for tasks which it alone used to perform, there may not be many changes on the ground. In fact, there may be new inequalities as established companies alone secure these contracts. Reform-minded individuals need to be posted in key positions of authority for substantial periods to inspire attitudinal change down the line. Officials need to develop negotiating skills and require model contracts for reference while dealing with private parties with high level of skill but low collateral. Thus, confronting the issues involved in making durable improvements in solid waste management and ensuring appropriate changes in policies and structures posed a daunting challenge to decision-makers.

Since the early 1990s there have been international and national developments that have inspired and encouraged innovative actions on the part of governments, committed individuals and citizen groups to improve existing waste management arrangements in Indian cities, including Delhi. The emergence of a new perspective on waste management linked with a revised view of 'development' that has found increasing acceptance in the world community has been of great significance in this connection. As a result, funding patterns of the World Bank and other international lending agencies that earlier focused mainly on rural areas in the developing world, in recent years, have broadened their interest to include improvements in urban infrastructure. In India, the notification of the Constitution (74th) Amendment Act in 1993 which was meant to energise local governments in urban areas has also invigorated civil society. Non-governmental organisations have begun to take an interest in waste related issues and the result has been a continuity of media attention on this subject that has no parallel in earlier periods.

In 1992, in Rio de Janeiro, at the World Conference on Environment and Development, India, like the other 171 countries present, pledged itself to observe the principles of 'sustainable development' in its economic activities. The core idea behind this concept was that benefits of development should be equitably distributed and environmental resources treated with restraint so that future generations are not at a disadvantage. India and the other countries also accepted an action plan called Agenda 21 to reorient existing practices. In the context of waste management this Agenda indicated that methods of collection, transportation and disposal should be protective of health and productive capacity of all waste workers, regardless of whether they were from the municipal workforce or the informal system, and that waste emission levels should be within the capacity of the local environment to absorb pollutants. Further, there should be encouragement to recover resources from waste. Since the heightened responsibilities required a higher outlay, local governments which had a primary responsibility in this connection would have to not only trim organisation costs but also extract a payment from the beneficiaries which could be linked with the ability to pay. Chapters 20 and 21 of Agenda 21 emphasised that broadly speaking the principles of reduce, reuse, recycle and safe disposal were equally applicable to all types of wastes—non-hazardous waste from homes, human waste which it was said should be treated as potentially hazardous waste, and also to toxic waste from factories. Government authorities—central, regional, and local—should coordinate their functions to apply these principles that were in effect waste management tools.

With the passage of the Seventy-fourth Constitutional Amendment in 1993, local governments in urban areas in India acquired a constitutional basis. In the case of Delhi, today, 134 directly elected members compose the deliberative wing of the Municipal Corporation of Delhi (MCD) whose jurisdiction extends to over 94 per cent of the National Capital Territory of Delhi. In addition, the councillors choose members of the 12 Ward Committees corresponding to the 12 zones into which MCD's jurisdiction is divided. The Municipal Commissioner, the head of the Executive is a senior civil servant. Thus legislative and administrative and even supervisory structures have recently been put in place by which policy initiatives designed to meet the requirements of 'felt' need at the grassroots level can be composed and implemented. The significance of these in impacting on the quality of life of ordinary citizens was made dramatically evident in February 2002 when the Bharatiya Janta Party (BJP) a new form of the defunct Bharatiya Jan Sangh Party which was born in Delhi and which controlled the MCD was ousted by the Congress party. It was widely believed that the BJP had lost because it had neglected to utilise new opportunities at the local level to expand its traditional base which consisted of migrants from Punjab in the late 1940s and early 1950s, to include the more recent entrants from Bihar and UP, who were by and large poor.

In 1991 yet another structural change, supportive of greater dynamism in waste management was ensured with the passage of the National Capital Territory of Delhi Act, 1991 which provided for a legislative assembly of 70 members. There had been a partially elected body earlier. The powers of the Legislative Assembly to which the Chief

Minister and the Council of Ministers were responsible were made similar to those of other state assemblies in the country, with some exceptions. From 1993 till 1998, BJP was in power in Delhi and in 1998, the Congress Party became the ruling party and retained its position in the next elections in December 2003.

Media focus in recent years has helped to generate debate and sustain interest in management options in relation to all four categories of waste. A major stimulus has been the Supreme Court's wide-ranging actions in response to Public Interest Litigations (PILs) filed by concerned citizens and groups. In the case of Dr. B.L. Wadhera *versus* the Union of India and others for instance, where attention was drawn to litter in public places, the Supreme Court in 1996, issued a 14 point directive relating to improvements in waste management to the MCD, the New Delhi Municipal Committee and also to the concerned departments of the Central government and Delhi government. Further, each of these bodies was asked to submit compliance reports separately to the Court. In addition, the Supreme Court appointed the Central Pollution Control Board (CPCB) as the sole monitoring agency and asked it to submit a report every two months. In response to the PIL lodged by the Centre for Science and Environment (CSE) and others in which it was alleged that the Yamuna river had become a virtual sewer, the Court indicated a deadline for installing specific pollution control measures including public toilets to ensure that no untreated wastewater was discharged into the Yamuna river. In 1997, in connection with the PIL filed by the Research Foundation for Science, Technology and Natural Resource Policy, New Delhi, challenging the import of hazardous waste into the country, the Supreme Court appointed a High Power Committee (HPC) whose scope of work as envisaged by the terms of reference was extended to produce a long term perspective document related not only to the regulation of import of hazardous wastes but also management and safe disposal of such waste generated in the country.

When the Congress Party formed the government in Delhi in December 1998, improvements in handling and disposal of wastes was a priority item on its agenda. One of the main objectives of her government, as the Chief Minister, Shiela Dikshit, subsequently said, was to make Delhi, "a centre of urban excellence, a well managed, clean and dynamic city." A 'participatory and interactive framework of governance' was considered indispensable for this objective.

The purpose of this study is to comprehend how the management of solid waste understood in the four dimensional holistic sense was undertaken during 1998-2003, the first tenure of the Shiela Dikshit led Congress Government of Delhi. 2003 was considered a convenient date for ending the study because popular perception of this government's skill in reconciling environmental and developmental concerns was regarded as a major factor which contributed to the victory of the Congress in Delhi, the only State in which the Party won when elections were held for four state Assemblies in November-December 2003. In waste management the role of the State government is limited and in Delhi's complex governing structure it is even more so, but nevertheless it is a decisive source of influence. The role of local government is most crucial. In Delhi there are three—New Delhi Municipal Corporation, Delhi Cantonment Board and the Municipal Corporation of Delhi. The activities of the MCD alone will be analysed because 94 per cent of Delhi is under its jurisdiction.

The waste management operations will be examined through the lens of the sustainable solid waste management (SSWM) perspective. Though the main ideas of this perspective offered little which was new to government functionaries familiar with liberal socialist principles, it will be assumed that concerned decision-makers in the MCD were familiar with implications of this perspective for improving waste management in Delhi due to their presumed association with a World Bank sponsored project which began in the late-nineties. This study initiated by the Ministry of Environment and Forests and the Delhi Government was meant to suggest policies and action plans to make Delhi a more environment friendly and better governed city by 2021. A panel of experts appointed by GK International and Operation Research Group, India, met with government officials and civil society members to review existing arrangements relating to various aspects of development and discussed various options before finalising the report which became available to the public in January 2001.

Objectives of the Study

The first objective of the study is to identify the socio-economic features of the city. The background and occupations of its citizenry, the land use patterns which impact on waste generation and the components of Delhi's government system that have a role in the management of waste will be noted. Features of a perspective on sustainable development for Delhi as developed by the Delhi Urban Environment and Infrastructure Project (DUEIP), known as Delhi-21 will also be indicated.

The next objective is to understand the nature of solid waste in a developing country as compared to a developed country. The features of SSWM paradigm as evident in the Brundtland Commission Report, Rio Principles and Agenda-21 will be highlighted and its differences with the concept of solid waste management (SWM), identified. In the assessment, the value of the perspective as a decision-making guide will be discussed.

The object of the next three chapters is to understand the manner in which four kinds of solid waste in Delhi non-hazardous waste and hazardous waste—further sub-divided into three, toxic chemical waste, biomedical waste and human waste have been managed in recent years. The following points will be considered:

- Problems posed by the waste due to its composition and quantum and its social and environmental impacts.
- Features of the Central government's initiatives in the collection, and safe disposal of the waste.
- Suggestions for progress towards its sustainable management as indicated in Delhi-21.
- Steps taken to improve management of the waste by state and sub-state authorities between 1998-2003.
- Assessment of the impact of these initiatives in the light of SSWM and indication of measures that could have been taken within existing budgetary and structural constraints to reinforce positive trends.

In the conclusion, it will be shown, how though in Delhi there has been considerable advance in the direction of SSWM in recent years, much more needs to be done in the context of 'issues,' 'policies' and 'structures' for Delhi to move into a leadership position in the country in this most significant aspect of public life.

Data/Sources

Both primary and secondary source material was used for this study. Meetings were held with spokespersons of three NGOs. Toxics Link, Asian Centre for Organisation Research and Development, and Centre of Environment Education together with officials of Conservancy and Sanitation Engineering Division of the MCD, Department of Environment and Bhagidari Cell of the Delhi government and with academicians in the Indian Institute of Public Administration and the Institute of Human Development, in New Delhi.

Relevant Acts and Reports of Committees that have significance for this study, such as the Barman Committee on Solid Waste Management in Class 1 cities of India (1999) and the High Power Committee on Control of Hazardous Waste (2001) in India were examined. The Rio Principles and Agenda-21 adopted at the United Nations Conference on Environment and Development and India's report on what it had done to implement it as contained in its report, Agenda-21: An Assessment, were studied. Delhi Urban Environment and Infrastructure Improvement Project Report, State of Environment: Report for Delhi 2001 and Delhi Government's Working Report, 4 Years (1998-2002), were extensively consulted.

Secondary sources, press clippings, journals and magazines were studied mainly in the libraries of the India Habitat Centre, New Delhi, World Bank, New Delhi, Indian Institute of Public Administration, New Delhi and United Nation's Economic and Social Commission for Asia and the Pacific, Bangkok.

In addition the Internet was used to access primary and secondary sources. Internet sites most frequently used were those of the Planning Department, Delhi government, Ministry of Environment and Forests of the Government of India, Central Pollution Control Board, Planning Commission of India, Census of India, Centre for Environmental Law Education, Research and Advocacy, World Bank and United Nations Department of Economic and Social Affairs, Division for Sustainable Development.

1

Delhi

Delhi, modern India's capital city, with a history dating back to the 15th century B.C. due to its identification with *Indraprastha* which finds a mention in *Mahabharata*, is located at a site from where the highest political authority has been exercised, off and on, since the 12th century A.D. Tomar Rajput rulers are credited with founding Delhi. The first regular defended system arose in 736 A.D. and was called *Lal Kot*. Five more cities followed before the Mughal Emperor, Shah Jahan, built Shahjahanabad, the seventh township, between 1639 and 1648, on the banks of the river Yamuna. The British took over Delhi in 1803 and decided to make it the capital of the British Empire. The eighth city, the present New Delhi where symbols of imperial architectural granduer are still intact, was built in 1911 when the capital was shifted from Calcutta. Contemporary Delhi, the ninth city as it were, is the end result of the vast changes that have occurred since Independence in its physical and social composition and also in its economic orientation through a pattern of growth which is frequently described as 'haphazard.'

Socio-Economic Aspects

Growth in Population

The population of Delhi grew from about 4,00,000 in 1901 to about 9,00,000 in 1941. With the huge influx of refugees, after the partition of India in 1947, the population grew between 1941-1951 by almost 90 per cent. Population growth in Delhi has been over 50 per cent in each successive decade since 1951. As per the first census of independent India, population of Delhi in 1951 was 17,44,072. Between 1951-1961, the growth in population was 52.44 per cent and from 1961-1971 it was 52.93 per cent. The increase was 53 per cent in the decade 1971-1981. Therc was a slight decline to 51.45 per cent in the growth rate between 1981-1991 and a further decline in the growth rate between 1991-2001 to 46.31 per cent.

Delhi's population is 13.783 million as per the 2001 census. Some of the main demographic features of Delhi that emerged in the 2001 census are:[1]

- Delhi's annual growth rate of population is on the high side. During the period 1991-2001, it has been 3.8 per cent, which is almost the double of the national average.
- Delhi's population density is also on the high side. At 9,294 persons per square kilometre as per the 2001 census, it is the highest in the country. In 1901, the density was 274 persons per square kilometre and by 1951 it had increased to 1,176 persons per square kilometre.
- Delhi has the highest percentage of urban population in India at 93.01 per cent.
- Delhi is the third most populated urban agglomerate in India.

Migration

The National Capital Region Plan had expected that the population of Delhi would be 11.2 million in 2001 and that 2 million migrants would be deflected to the counter magnet towns in the National Capital Region Area by relocation of economic, industrial, and governmental activities. This plan proved to be ineffective. Delhi's population is expected to become 19.5 million by 2011 and 22.4 million by 2021, as the trend of migration of increasing numbers is expected to persist.

The decline in the decennial growth in population during the last two decades has been more than made up by the net increase in migration which had been steadily increasing since 1961 and became 'more than one-and-a-half times, the natural growth in population (1:1.55) in 1997.' In absolute terms, the natural increase in population during 1991 was 2.11 lakh compared to migration of 1.78 lakh. The natural increase reached 2.18 lakh in 1997 but the migration was estimated at 3.37 lakh.

---

1. *Economic Survey of Delhi*, 2001-2002, pp. 14-25, *www.delhiplanning.net.in*.

Estimates of migration to Delhi from 1991 to 1997 are given in Table 1.1 below.[2]

Table 1.1

Estimates of Migration–Delhi

(in lakh)

Year	*Population as on 1st July*	*Increase in Population*	*Natural Increase*	*Increase due to Migration*	*Ratio (Natural vs. Migration)*
1991	95.60	3.89	2.11	1.78	1: 0.84
1992	99.89	4.29	2.12	2.17	1: 1.02
1993	104.38	4.49	2.06	2.43	1: 1.18
1994	109.07	4.69	1.94	2.75	1: 1.42
1995	113.97	4.90	2.06	2.84	1: 1.38
1996	119.15	5.18	2.07	3.11	1: 1.50
1997	124.70	5.55	2.18	3.37	1: 1.55

Source: 1. Population–Registrar General and Census Commissioner of India.
2. *Annual Report 1997: Registration of Births and Deaths*, Directorate of Economics and Statistics, Delhi Government.

The character of migratory population, as recorded in various studies, reflects the following characteristics:

- Rural to urban migration is the most dominant type, and among them, life long migrants (long staying) constitute two-fifth (40.20 per cent) of Delhi's migrating population in 1991. Migrants from Bihar, Haryana, Punjab, Rajasthan and Uttar Pradesh account for 85 per cent of total migrants.
- The employment structure of the migrated slum-dwellers in Delhi is by and large dominated by tertiary-sector occupation. The percentage of migrants below the poverty line is highest among those who migrated relatively recently (between one month and five years) and this percentage declines with the period of migration.
- 64 per cent of male migrants moved for economic reasons and 20 per cent for family reasons. Males were between 15 and 39 years and among females the dominant age group was 20 to 34 years. Most migrants were illiterate or poorly educated.[3]

---

2. *Economic Survey of Delhi*, 1999-2000, *www.delhiplanning.net.in*
3. The Energy Research Institute (2001). *State of Environment Report for Delhi 2001*, pp. 13-16, Delhi.

Occupation Profile

Migration has been spurred by the fact that Delhi evolved from being the hub of politics and administration to a multi-functional city with large scale industrial, commercial and service-oriented activities. In the post-Independence period, Delhi "emerged as a major industrial and commercial sector with phenomenal growth in manufacturing, trading, communications and transport activities" to become the most prosperous among the States and Union Territories of India.[4] Major manufacturing units in Delhi are concerned with food, metal, textiles, paper and paper products and repair services. The manufacturing sector served as the growth engine for related activities such as transport, trade, commerce and construction. The economic liberalisation process has provided a further stimulant to these sectors.

Investments have favoured the growth of small-scale manufacturing. Percentage of people engaged in the industrial sector increased from 16.94 per cent in 1951 to 32.43 per cent in 1991 and the rate of increase of the workforce during 1981-1991 was 48.85 per cent compared to 28.42 per cent at the national level during the same period. The Gross State Domestic Product (GSDP) of Delhi was Rs. 65,657 crore in 2001-2002. While Delhi's population is 1.34 per cent of the country, the share of Delhi in country's GDP is 3 per cent. The contribution of the tertiary sector to the GDP of the State economy grew from 68.5 per cent to about 81 per cent between 1993-1994 and 2001-2002. The per capita income of Delhi at Rs. 42,818 was about 2.4 times that of the national average of Rs. 17,978 in 2000-2001. It is therefore not surprising that Delhi attracted a large number of migrants resulting in a demographic growth rate during the 1990s higher than that of most metro cities.[5]

According to the Economic Census prepared by the Delhi Directorate of Economics and Statistics on the basis of Economic Census conducted in 1998, the number of enterprises grew by 5.27 per cent per year during 1990-1998 compared to the growth of 4.60 per cent during 1980-1990. By the end of the Eighth Five Year Plan in 1997 there were 1.26 lakh industrial units. Nearly 99 per cent of

---

4. Kundu, Amitabh, (2003). "Population Dynamics and Economic Development in Delhi," Seminar on *Sustainable Development of Delhi*, June 16-17, p. 7, New Delhi.

5. *Ibid*, p. 6.

the total enterprises were managed by the private sector. Out of the total enterprises only four per cent were found registered under various authorities. The annual growth rate of employment in urban Delhi was 6.6 per cent during 1990-1998, almost double of the annual growth rate of 1980-90. Out of the 35 lakh people employed, around 2.83 lakh or 8.08 per cent of the total number were women. Delhi rated eighth in terms of employment in comparison with other states. The survey also revealed the strong trading character of the city with 33.7 per cent of the enterprises managed by traders. The social, community and personal services, including professionals, constitued 21.52 per cent. The share of the manufacturing sector was 18.86 per cent.[6]

Socio-Economic Divisions

In 2003, *Hindustan Times* (HT) and the Centre for Study of Developing Societies (CSDS) carried out a survey across 70 assembly constituencies in which over 14,460 persons were interviewed on a wide range of subjects. On the basis of the data gathered, a class profile of the city was generated. Six categories of people were made from 'very rich' to 'very poor' primarily on the basis of assets owned. As per the classification, the very rich in Delhi constitute those that have two or more cars or two or more air conditioners. The rich in Delhi have two of the following—car, air conditioner, computer and telephone. At the other extreme are the very poor, those with no material possession. Just above the very poor in the economic ladder are the poor who live in a *jhuggi-jhopri (jj)* or a *kutcha pucca* mix house, get drinking water from a MCD hand pump or from a MCD tanker and possess either a TV or a cycle or a radio. In between these classes are those classified as middle class. They live in a Higher Income Group (HIG) or Middle Income Group (MIG) flat, have a car or a scooter, a colour TV and a refrigerator and at least one telephone. The lower middle class in Delhi lives in a *kutcha pucca* mix house, either a Janta or a Lower Income Group flat and possesses two of the following—scooter, refrigerator, telephone, colour or black and white TV.

According to the survey, six per cent of the people living in Delhi are very poor and yet another 28 per cent are poor. If clubbed

6. *Fourth Economic Census of Delhi* (1998). Directorate of Economics and Statistics, Government of National Capital Territory of Delhi, New Delhi. *www.delhigovt.nic.in*

together, 34 per cent can be called poor. On the other hand nearly six per cent can be considered very rich and 12 per cent as rich. As such 18 per cent can be said to be in the rich category. There is a big middle class. Out of the 48 per cent that comprise this category 28 per cent belong to the upper-middle class and 20 per cent are in the lower-middle class. If the percentage figures relating to the relatively poor are added, that is 6, 28 and 20, then clearly it is the poor who constitute the majority in Delhi. The HT-CSDS survey shows that this is at present the most politically significant section of the population. In December 2003 elections while 53 and 54 per cent of the poor and very poor voted for the Congress only 24 and 18 per cent of these categories voted for the BJP. Class cut through caste in a city where due to changes in the pattern of migration, the post-Independence political elite which was drawn largely from Punjabi *Khatris*, *Jats* and *Brahmins* has broadened to include members of the other backward class (OBC) and *Dalit* communities who shifted base from Bihar and UP.[7] In December 2003 elections, the Congress won in Delhi because it was viewed as the party committed to encouraging mass level rather than merely elite level consumption. Since generation of waste is a function of production-consumption to which every body contributes, the social profile of Delhi suggests that Delhi's authorities should give high priority to sound waste management if they want Delhi to develop into a world class mega-city comfortable about coping with its internal socio-economic contradiction.

Urbanisation of Delhi

Another dominant trend in the growth of Delhi has been its increasing urbanisation. As per 2001 census out of the total population of 13.783 million, 12.82 million is urban and .963 million is rural. The urbanisation of Delhi dates to the beginning of the 20th century. In 1901, 52.76 per cent of the total population of Delhi was urban and 47.24 per cent was rural. The rural population continued to decline till 1981 when it became 7.27 per cent of the total population. It temporarily increased to 10.07 per cent in 1991. This reversal of the long-term trend during 1981-1991 was due to mushrooming of unauthorised colonies in rural areas. In the decade

---

7. Kumar, Sanjay (2004). "A Tale of Three Cities," in Seminar, 534, *Beyond the Verdict*, February, Delhi.

1991-2001, the population growth reverted back to its original pattern, with the rural population declining to 6.9 per cent of the total population. The urban area in Delhi territory has increased from 22 per cent in 1961 to 62.5 per cent of the total area in 2001. In the total area of National Capital Territory (NCT) Delhi of 1483 sq. km., the urban area has grown from 526.26 sq. km. in 1971 to 891.09 sq. km. in 2001. The number of rural villages has declined from 300 in 1961 to 165 in 2001.

Table 1.2 shows the transition in rural urban population from 1901 till 2001.[8]

Table 1.2

Rural-Urban Area and Rural-Urban Population Percentage

Year	*Area sq. km. Rural*	*Area sq. km. Urban*	*Rural Population*	*Urban Population*
1901	—	—	47.24	52.76
1951	—	—	17.60	82.40
1971	891.20	593.80*	10.30	89.70
1981	891.10	591.90	7.27	92.73
1991	797.66	685.34	10.07	89.93
2001	591.91	891.09	6.9	93.01

* In 1971 census, Delhi's area was indicated as 1485 sq. km.

Source: General Population Tables, Registrar General and Census Commissioner of India.

An analysis of the Survey of India maps published in 1996 show that 645 square kilometres constitute the built-up area in Delhi comprising 44 per cent of the geographical area. This includes Delhi's urban limits, built-up area in rural zones and farmhouses. The urban sprawl is more towards eastern and southern part of Delhi. The maps reflect the perception of the planners over the years that to accommodate the growing population, not only was there need to increase the population holding capacity of the area in the urbanised limits there should also be its further extension into rural use zones. Physiographically, much of Delhi is covered by the fertile Yamuna alluvial soil. Urban expansion has occurred both at the expense of irrigable agricultural land with its range of fine-to-coarse loamy soil-ideal for cereals and pulses, in not only rural-use zones but also in spaces earmarked for recreational purposes such as parks, open areas

8. *Economic Survey of Delhi*, 2000-2001, *op. cit*, Chapter III.

and green areas. Though the Master Plan had envisaged that each planning area should have only one use, in practice there has been a mix of commercial and residential use. Thus, the urbanisation process has been not only extensive but also haphazard.[9]

With rapid urbanisation, Delhi has progressively lost its green cover. In 2001, it had a mere 88 square kilometres of forest cover in the total geographical area of 1483 square kilometres. The tree planting programmes have not been very successful because of the very low survival rate of trees in Delhi due to lack of proper maintenance. It has been estimated that "considering earlier records on various plantation schemes that had been undertaken since 1950, Delhi should now be having 470 sq km of dense forest cover even after accounting a low survival rate of 5 per cent-6 per cent."[10]

The natural ridge forest which has served as the lungs of Delhi has also dwindled considerably, mainly due to human interventions. The degradation of the ridge area which started way back in the 1920s continued until recently due to various encroachments for constructing buildings, roads, settlements, parks, etc. Large chunks of area are also being used for activities such as garbage dumping, apart from fuel extraction and fodder collection and grazing by livestock. The shrinking of ridge area has caused enormous loss to its bio-diversity especially the loss of species of flora and fauna. Many concerned citizens and NGOs have joined hands to combat this environmental menace and prevent further degradation of the ridge forests. Mainly due to their efforts, a large portion of the ridge was declared as a 'protected' zone in 1980. Subsequently, many encroachments were removed because of the interventions of the Supreme Court. Large chunks of land on the ridge have also been notified as reserve forests.[11]

Habitats in Delhi

The era of urban planning in Delhi commenced in 1824 when a Town Duties Committee was constituted for the development of the Cantonment, Khyber Pass, the Ridge area and the Civil Lines area by the British. Urban development picked up in Delhi with the

9. *State of Environment Report for Delhi 2001, op. cit.* p. 6.

10. *Ibid*, p. 78.

11. *Ibid*, p. 95.

composition of the first Municipal Committee of Delhi (MCD) in 1883 and the Municipal Building bye-laws were first made operational in Sadar Bazar, Subzi Mandi and other suburbs. The Delhi Improvement Trust (DIT) was constituted in 1937 for the planned development of the city.

In 1957, the Delhi Development Authority (DDA) was set up under the Delhi Development Act, 1957, as a successor to the DIT for the planned development of Delhi. The DDA prepared the first Master Plan for Delhi to ensure appropriate balance between the spatial allocations for the distribution of housing, employment, social infrastructure, and transport, and adequate arrangement to accommodate all other physical infrastructure and public utility systems in Delhi. The Master Plan for Delhi (MPD) 1961-81, with a view to ensure appropriate allocation of land and development of all public utility/physical infrastructure, made provisions of nine land-use categories with further 37 use zones.

The first Master Plan of Delhi (1961-1981) envisaged development of 44,770 hectares of urban area by 1981 for urban population of 41,00,000 persons. Subsequently, development of an additional area of 4,000 hectares of urban area was added to the target of the first Master Plan. The first Master Plan was reviewed and amended to provide for its extension for another 20 years (1981-2001) and was published in 1990. The amended Master Plan (Second Master Plan 2001) envisaged acquisition of another 20,000 hectares of land for urban area extension of Delhi by 2001 and set a target for development of 68,770 hectares urban land. This was subsequently proposed to be enhanced to 83,804 hectares of land as urban area within the framework of MPD 2001.

By 1998, the DDA had facilitated the construction of about one million dwellings under various schemes. These included about 258,230 units constructed by the DDA for general public and about 240,000 dwellings in resettlement colonies for squatters. About 525,000 dwellings were generated through leasing of land by the DDA for group housing and cooperative group housing societies.[12] However, by and large, the objectives of the Master Plan could not be achieved.

---

12. Risbud, Neelima (2002). "Policies of Tenure Security in Delhi", in A. Durand-Lasserve and L. Royston (eds), *Holding Their Ground: Secure Land Tenure for the Urban Poor in Developing Countries*, p. 60, Earthscan, London.

The measures taken over the years by the DDA proved to be inadequate to meet the growing demand of land to meet the basic need of shelter. It could not provide enough developed land at affordable prices to different categories of residents. It was also unable to make available enough developed land for physical infrastructure and public utilities to the concerned agencies in time. This failure can be mainly attributed to the fact that it was decided to acquire land in bulk and the DDA, an autonomous body of the Central government, alone vested to manage, develop and dispose it only on leasehold basis, created in the process, a government monopoly in land.

Informal 'land delivery sub-systems' evolved and expanded to satisfy this need and in the process outstripped the public supply system. Private initiative, which was not part of the official land development policy, took the form of squatter settlements, unauthorised colonies and misuse of urban villages. These differed from each other in terms of process of development, tenure, the nature of illegality and the resultant settlement characteristics. These areas were not entitled to a standard level of services owing to their illegal/quasi-illegal status.[13]

Squatter Settlements

Squatter settlements were basically encroachments upon public land by the migrant poor. Tenure was illegal and perceived security was variable. The steady rise in the number of poor migrants into the city has meant the emergence of squatter settlements in all parts of the city including large sized clusters in prime central locations such as Minto Road, Gole Market, Matia Mahal and medium sized clusters in posh Delhi colonies such as Hauz Khas and R.K. Puram. The larger sized squatter clusters in the three central locations contributed half the *jhuggi jhopri* (JJ) population of the city in 1997.[14]

Programmes for resettling squatters, with varying tenure conditions and standard of development, have been implemented since 1960. The first public policy for squatter settlements was proclaimed by the Central government in 1960 when it was decided

---

13. *Delhi-21*, Delhi Urban Environment and Infrastructure Improvement Project (DUEIIP) Report, (January 2001), Government of India, and Government of National Capital Territory of Delhi, p. 26.

14. *State of Environment Report for Delhi*, 2001, *op. cit.*, p. 16.

to remove the squatters from public land and resettle them in planned areas under the *jhuggi jhopri* removal scheme. Those squatting prior to 1960 were considered eligible for alternative accommodation but those squatting after 1960 were to be removed. The scheme was revised many times in the 1960s and 1970s and a large-scale clearance and relocation of squatters took place in the 1970s without "conforming to land use stipulation of the Master Plan with regard to zoning regulation, plot sizes, building by-laws or municipal regulation of water supply and sewerage."[15] By mid-1980s, about 2.40 lakh JJ families had been resettled in 46 resettlement colonies where these families were given plots between 21 square meters to 80 square meters with basic infrastructure facilities provided in phases.[16] Leasehold rights over the plots allotted were given to the residents that entitled them to a higher level of civic amenities.

In 1990, it was decided to give ration cards (food cards) to all *Jhuggi* dwellers. This led to the first comprehensive authentic survey of *Jhuggi* dwellers that led to the identification of 929 JJ Clusters with a population of 2.60 lakh families. Since 1990, a three-pronged strategy is being followed by the Delhi government to deal with the problem of *Jhuggi-Jhopris*, in line with the Central government's policy on JJ dwellers. Within the broad framework of not permitting any fresh encroachments, the encroachments in existence prior to 31 January 1990 were not to be removed without providing alternatives. The three strategies were:

- Relocation of JJ colonies on case to case basis where the original land owning agencies urgently require the encroached land for priority public purpose projects and were willing to contribute the entire cost of developed site in the new location.
- *In-situ* upgradation of JJ colonies where the land owning agencies have no plans for the use of the encroached land and give a 'No Objection Certificate, for its development for the JJ dwellers.' One essential feature of this scheme is the construction of Pay and Use *Jansuvidha* Complexes containing toilets and baths.
- Provision of basic minimum civic amenities for the community use under the scheme of Environmental improvement of Slum

15. Risbud, Neelima, *op. cit.*, pp. 61-62.
16. *Tenth Five Year Plan of Delhi, Urban Development*, Planning Department, Government of Delhi, pp. 1420, 1422, *www.delhiplanning.nic.in*

and JJ clusters, including provision of toilets and baths under its component scheme of construction of Pay and Use *Jansuvidha* Complexes containing toilets and baths, irrespective of the status of the encroached land until its coverage under one or the other of the above two schemes.[17]

This scheme was modified from time to time in the 1990s; the cut off date for eligibility for allotment of land was extended from pre-31 January 1990 to pre-31 December 1998 squatter families; the land was now given on license basis; plots of land measuring 18 square meters was given to pre-31 December 1990 families and of 12.5 square meters to post-31 December 1990 families; and the provision of plinth and WC was given up in the future development of plots.

Pursuant to a directive by the Delhi High Court in November 2002 to reconsider the policy of relocation of slums, a coordinating committee set up to consider this issue, broadly recommended allotment of a uniform plot size of 18 square meters to pre-1990 as well as post-1990 squatter families up to December 1998 and provision of essential services such as transport, education, health-care, fair price shop, electricity, water, sewerage, approach roads before or simultaneously with the shifting of the JJ families to the new settlement sites. Keeping in view the magnitude of the ever-increasing problem of *Jhuggis, in-situ* upgradation of *jhuggis* was to be encouraged.[18]

The purpose of the scheme of resettlement was defeated to some extent because of the unauthorised sale of tenure to higher income groups as had earlier happened in the case of plots of land allotted in the earlier resettlement colonies. The scheme of *in-situ* upgradation of JJ Clusters, which involved reorganisation of the slum-dwellers on the same site with a planned layout, licensed cooperative tenure and minimum basic services to enable families to construct their own houses, had also not been very successful because all the conditions of the licensed tenure could not prevent unauthorised selling of plots to higher income families. Besides, the majority of land owning agencies refused to give permission for construction of facilities within these clusters. It is also debatable whether the *in-situ* resettlement could have worked given the vast number of residents involved. Under

17. *Ibid*, pp. 1423-1424.
18. *Ibid*, p. 1433.

this strategy, since its inception, 784 families have been covered in the three schemes at Ekta Vihar, Prayog Vihar and Madrasi Basti.[19] Judicial pronouncements since the 1990s have stalled attempts to stabilise slum rehabilitation through security of tenure.

The Delhi High Court in January 2003 struck down the Government's slum rehabilitation policy. The court order implies that the slum dwellers would no longer be able to get plots in lieu of their *jhuggis*. The High Court order stated that, "the Government bears a substantial cost of slum rehabilitation which is illegal as it amounts to buying back the encroached public land from the *jhuggi* dwellers. First, the land is acquired using public money, making the farmers landless. Then it is allowed to be encroached upon, and again cleared at public cost." The City government subsequently filed a Special Leave Application against the High Court order in the Supreme Court.[20]

Unauthorised Colonies

Unauthorised colonies were areas where due to prior existing condition of illegal sub-division of land and built-up areas on them, the DDA could not complete the formality of acquisition of such land for development and subsequent development. Before the DDA came in to existence, land for settlement was developed either by the group cooperative housing societies for its members or by private colonisers for subsequent sale. In both cases, land was developed according to layout plans duly approved by the government authorities. However, along with the development of such planned colonies, unauthorised colonies were also coming up in Delhi without approved layout plans by illegal sub-division of land to cater to the needs of those who could not afford to purchase land from the private colonisers or pay the development charges asked for by the housing cooperative societies. The sub-division was illegal either because it violated zoning regulations or sub-division regulations. Because of the illegality of sub-division, such colonies were not eligible for provision of infrastructure services and therefore lacked basic civic amenities. By the time the DDA came into existence, there were already 110 such unauthorised colonies with a population of 221,000. When the DDA started its

19. *Ibid*, p. 1439.

20. Sharma, Aruna P. (2003). "HC Order Blocks Shifting of Slums," *Hindustan Times*, January 8, Delhi Edition.

acquisition of land, it was decided to exempt such built-up areas from acquisition and to regularise them if they could be fitted into 'the sanctioned regularisation plan.' This policy called for the regularisation of all 110 unauthorised colonies though some of them could not be regularised inspite of Government's decision to regularise them because 62 of these colonies were located on non-conforming land uses as per Delhi's Master Plan.[21]

Failure of the DDA to develop adequate number of colonies and provide sufficient number of plots to meet the growing need of the people in the 1960s and 1970s led to the growth of more unauthorised colonies inspite of the fact that the Delhi Lands (Restriction on Transfer) Act was adopted in 1972 to stop the sale and purchase of land without permission of the Competent Authority. By the middle of 1977, 607 unauthorised colonies had come up within the jurisdiction of the MCD.

Government policy towards unauthorised colonies has been to extend regularisation to increasing number of properties within the framework of the Master Plan in return for payment of development charges by modifying its original conditions. Thus, in 1977 the Central government announced that all residential and commercial structures within the union territory of Delhi, which arose out of illegal subdivision of land till June 1977, irrespective of their date of origin, should be regularised including those located in designated slum areas. Master Plan would be changed wherever necessary for the regularisation of these colonies even as alternative accommodation would be given to those who would be displaced in the process of providing roads and facilities.

The residents considered the regularisation standards too high. Out of 607 colonies, regularisation process was started in 543 colonies but it made very slow progress. Recovery of development charges was poor and most people stopped paying after the first instalment. Anomalies between the actions of different institutions created ambiguities about the status of settlement and raised hopes of its survival where the stated regularisation regime was being defied and no action was taken to enforce it. According to the Delhi Municipal Act, development without approved layout was punishable with imprisonment. Plots

21. Risbud, Neelima, *op. cit.*, pp. 67-69.

earmarked for facilities in the regularisation plan were invariably encroached upon or sold by the colonisers. Construction work proceeded even as basic municipal services were lacking. To meet this need through subsidised facilities pressure was maintained by the residents on the concerned politicians.

The growth of unauthorised colonies continued in the 1980s and early 1990s. Private developers were quick to identify an opportunity in a situation where a plot of land in an authorised colony could command a much higher price once the colony was authorised or regularised. In 1994, the Delhi government asked members of the Legislative Assembly to submit plans of the unauthorised colonies within their constituencies and residents were publicly requested to submit layouts of their colonies. Only a few layouts were received. The City government recommended the regularisation of 1073 colonies which had arisen before 31 March, 1994 on the basis of an aerial survey. The regularisation of the colonies was stalled because investigations showed a mismatch between the indication on the aerial survey and the physical check out.

In response to a petition against the regularisation move in which the NGO, Common Cause, had urged that the public authorities by their indifference were acquiescing in conditions that were degrading common facilities, the Delhi High Court issued a directive in 1997 to the Central government to announce a fresh clear cut off date for regularisation of colonies as also amount of development charges to be collected. Penalty in the form of demolition was to be against entire colonies that would not be regularised, not merely a few select houses. The Court further directed that until the matter of regularisation was decided, no construction of any nature be undertaken in any unauthorised colony of Delhi. In a departure from the earlier policy, regularisation of unauthorised colonies was to exclude a number of categories such as colonies in notified or forest areas, which would obstruct provision of infrastructure such as the metro rail and other utility networks, where the built-up area exceeded prescription of the Master Plan and where buildings had non-residential use with areas of more than 50 square meters. Further growth of unauthorised colonies was to be prevented by providing for prosecution of illegal colonisers by amending the Municipal Act and the DDA Act. The development charges to be recovered were much lower than the actual estimates for provision of services but not as low as the affected politicians were pitching for. In 2002 there were 567 regularised unauthorised colonies and 1017 unauthorised colonies. Many of the unauthorised colonies were on private land, some were on government land and a few were on land notified for acquisition where the DDA had already paid the compensation.[22]

Urban Villages

Urban villages are those rural settlements that have been engulfed by urban developments. These settlements are organic and therefore not illegal. Under the Delhi Municipal Corporation Act of 1957, rural settlements that become part of urban areas can be declared as urban villages. Rural settlements have been notified as urban in 1963, 1966, 1982 and 1994. As soon as the village is notified as urban, the *Panchayat* is superseded and the DDA becomes responsible for development of the urban village. After the development work is completed, it is transferred to the MCD for maintenance and upkeep.

Under the influence of market forces, these settlements have been transformed into residential and commercial complexes. The growing demand for low cost housing options is being met partly by the urban villages. There is also high demand by the affluent sections for construction of commercial complexes because building bye-laws though applicable are rarely enforced.[23]

Lack of land use regulations and MCD's permission in 1959 to set up cottage industries and to allow power connections resulted in indiscriminate and haphazard growth of industries, particularly small-scale polluting factories and overall 'haphazard and chaotic development' of the urban villages. Following the Delhi government's ban on polluting industries, several of them quietly operate in urban villages 'behind closed doors.'[24]

Till the end of 1970s, the development process largely ignored the urban villages. Residential, industrial and commercial complexes were planned around them. The DDA prepared separate development plans for these village settlements but the development process was extremely slow. In 1983, the Central government decided to provide basic amenities to the 90 urban villages that existed in Delhi then. An

22. *Ibid*, pp. 69-70.
23. *Ibid*, p. 64.
24. Aggarwal, Manu (2003). "City Backyards," *Down to Earth*, p. 56 June, 15.

attempt was made to use the opportunity to restrict the growth of industries in the urban villages by extending water supply, electricity and sewerage facilities to residents only. Though this did not prevent the setting up of obnoxious and hazardous industries, basic services became available in these areas to the residents. As in the case of unauthorised colonies, proliferation of industrial and commercial activities is very high in the urban villages.[25] The number of urban villages has now risen to 126 and it is the stated policy of Delhi government to provide all civic facilities to these urban villages.

Planned Colonies

With a view to increase the housing stock in planned colonies of Delhi, new realities were acknowledged. Power of attorney sales were accepted. Conversion of leasehold properties into freehold properties on payment of a fee was permitted and regularisation of unauthorised construction in legally approved buildings on payment of prescribed regularisation charges was given. Thus, there was a compromise between planned development and rising demand for housing stock.

Population in Different Settlements

According to an estimate, the population in different types of settlements in Delhi in 2000 is as under. (Table 1.3).[26]

Table 1.3

Type of Settlement and Approximated Population (2000)

Sl.No.	*Type of Settlement*	*Population in Lakh*
1	J.J. Clusters	20.72
2	Slum Designated Areas	26.64
3	Unauthorised Colonies	7.40
4	Resettlement Colonies	17.76
5	Rural Villages	7.40
6	Regularised Unauthorised Colonies	17.76
7	Urban Villages	8.88
8	Planned Colonies	33.08

Source: DUEIIP 2001.

25. Risbud, Neelima, *op. cit.*, pp. 67-69.

26. *Economic Survey of Delhi*, 2001-2002, *delhiplanning.nic.in*

Industrial Areas

Delhi was among the 12 worst polluted cities (air pollution) in the world according to the 1992 WHO/UNEP report on Urban Air Pollution in mega cities of the world. Industrial activity plays a vital role in the economic development of the National Capital Territory of Delhi. It contributes to about 21 per cent of the State Income of the Territory. Industrial growth in Delhi is a post-Independence phenomenon. Before the start of the planning era in 1951, there were 8,160 industrial units with a capital investment of Rs. 19.13 crore and a turnover of Rs. 35.35 crore employing 69,266 workers. According to the Socio-Economic profile of Delhi 2001-2002 brought out by the Delhi government, at the end of the Eighth Five Year Plan in 1997, there were 1.26 lakh industrial units employing 11.36 lakh workers with annual production of Rs. 6310 crore and investment of Rs. 2524 crore. As on 31st March 2002, the number of industrial units is reported to have increased to 1.29 lakh with employment of 14.40 lakh workers. This constitutes 41.4 per cent of the total work force in Delhi. The majority of these industrial units are located in non-conforming zones. The industrial growth is mainly in the small-scale sector as there are only a few industries in the medium/large sector.

When Delhi's first Master Plan was prepared, there were 8,160 industrial units in Delhi occupying an area of 1,000 acre. There were only two planned industrial districts and most industries were scattered all over the city and required relocation. The Master Plan supported industrial development in the interest of balanced economic development but considered it unwise to locate large and heavy industries in Delhi and categorised them as undesirable or prohibited industries.

For other types of industries, the Master Plan used techniques like industrial zoning. The Plan set aside 5,761 acre of land (modified in 1974 to 6,500 acre) for industries within the urban limits proposed for 1981. The bulk of land reserved for industrial use was in the form of industrial estates. Besides these, the Plan proposed flatted factories-multi-storied structures near residential areas for small-scale, non-nuisance industries. While most existing industries were permitted to shift to any industrial area, noxious ones could only shift to extensive industrial zones (designed to handle higher levels of pollution).

The Master Plan stated that the relocation process must be largely governed by the fact that there should be the 'minimum amount of dislocation of production' and the workers should not be put to 'undue hardship.' It laid down a time schedule that distinguished between noxious, nuisance and non-nuisance industries and allowed extensions to industries with higher capital value, more workers or more floor space per worker. For relocation of small industries not suited to flatted factories, but unlikely to be able to afford industrial plots, it said government should develop industrial estates where space could be rented out to them. In the next two decades since the Master Plan came into force in 1961, total industrial area of about 2,148 acre was developed and 3,300 units relocated. In 1981, there were 46,000 industrial units, the majority of them in areas not reserved for industries.

As per the revised Master Plan for Delhi approved in 1990, industries already 'prohibited' in Delhi as per the 1962 Plan, were again prohibited and ones that had continued or come up in spite of being prohibited were given three years to shift. The list of industries 'prohibited' in the 1962 Master Plan was expanded and included under the title 'Hazardous and Noxious Industries' as Section A in the H-category. The types of industries already identified as 'undesirable' in 1962 were listed in Section B in the H-category under the title of 'Heavy and Large Industries.' No new industries of these types or of the 81 types of 'Extensive Industries' included in the F-category were henceforth to be allowed in Delhi. Existing Heavy and Large Industries were to be shifted out according to the plan and policy for the National Capital Region. Existing F-category industries in areas not meant for them were to be relocated in existing and new extensive industry zones. In effect, only light and service industries were permitted in Delhi in the future.

Delhi's revised Master Plan approved in 1990 anticipated about 100,000 (mainly light and service–L and S) industrial units by 2001 and provided for well over 2,000 hectares of space for them. Sixteen L and S estates spread over 1,533 hectares and two extensive industrial estates over 265 hectares were proposed in the three urban extensions (Dwarka, Narela and Rohini). About 200 hectares of space in flatted factories, proposed in 1962, but not developed was again proposed. More space thinly spread across the city was envisaged through restricted industrial use in residential and commercial areas. 70 types of household industries were permitted subject to restrictions to protect residential area amenity. Another 100 types were permitted in urban villages. Nearly 150 types of industries were similarly permitted in planned commercial centres. Failure to implement 1962 Plan provisions had resulted in most units being in non-industrial areas and for these the revised plan proposed redevelopment or relocation. Redevelopment/regularisation was proposed for Shahdara, Anand Parbat and Samaipur Badli. Non-conforming F-category units and Light and Service units employing more than 20 workers were to shift to industrial areas within three years of being allotted plots.

Practically nothing was done to implement these plan provisions till the Supreme Court in a series of judgments from 1996 to 2000 intervened in the matter and gave extensive orders for relocation of industries. The Supreme Court, making the Master Plan of Delhi 2001 as a basis for its judgement, in series of orders given between 8 July 1996 and 26 November 1996, directed that 168 hazardous industries be moved from Delhi to suitable locations in the National Capital Region (NCR), or closed down by November 30, 1996; 513 and 337 extensive industries in residential and non-conforming areas respectively (H category in MPD-2001) be closed down by January 31, 1997, and relocated outside Delhi; 46 hot mix plants be closed down by February 28, 1997; 243 brick kilns be closed by June 30, 1997, and relocated outside Delhi; and 21 arc/induction furnaces be closed by March 31, 1997. The Court also directed the Delhi administration to ensure that all non-extensive F-category units operating in non-conforming areas, except light and service industries as permitted by MPD-2001, should be either relocated or closed down. In response to the Court's directive, a High Power Committee was required to examine the permissibility of functioning of units in the residential/non-conforming areas in terms of provisions contained in MPD-2001. All those industries, which were unable to obtain necessary permission from the High Power Committee, were to stop operating from residential areas with effect from 1st January 1997.

The Court continued to enter, cease and relocate order against non-complying and new industries till 2000. In November 2000, when the Supreme Court threatened to hold the Delhi government in contempt for licensing 15,000 industrial units in residential areas in violation of the Court's earlier orders, the Delhi government moved to close down the polluting factories while espousing the cause of regularisation of some of the industries.

Structures in Delhi's Governance

Three levels of structures, central, state and local are involved in governing Delhi. Their compositions, powers and mutual relations as they exist today have been shaped by a combination of historical circumstances that go back to before Independence and to the more recent demands of a developing democratic society. Their jurisdictions often overlap, perhaps more so than in any other city in India. "What makes Delhi city management peculiar and distinct is that the city's affairs are handled by a multiplicity of institutions."[27] Compulsions of competitive party politics, which can be intense, considering that both the major national political parties, the Indian National Congress and the Bharatiya Janta Party (BJP) are strongly entrenched in Delhi, are a major factor in moulding the decisions of public authorities. From the vantage point of the citizen at the receiving end of services the State is meant to provide, these features of the city create a very unsatisfactory situation. They also create difficulties for the authority figure most easily identifiable with the city. The Chief Minister of Delhi frequently lamented during her first term how she had to face brickbats for situations over which she had no control. In an interview in September 2001, Sheila Dikshit said, "All responsibilities of an elected government are perceived to be ours. In fact, we are held responsible for problems which are not under our purview. The Delhi government is sandwiched between the Union government and the civic bodies. We do not have any rights... and those who are actually responsible escape public criticism. The issue of relocation of polluting industries to conforming areas and the ongoing CNG crisis are some examples. People do not realise that our government is crippled with limited powers."[28] As has been often recorded in the media, a Delhi resident feels confused about deciding who should be held accountable

27. Pinto, M. (2002). *Metropolitan City Government in India*, p. 152, Sage Publications, New Delhi.

28. Singh, Shivani (2001). "Seeking Order Amidst Administrative Chaos," *The Times of India*, September 7, Delhi Edition.

for a lapse in public services to ensure a more efficient discharge of civic responsibilities. There is a strong case for review of the existing system of political and legal control.

Delhi Government

In India's federal polity, the counterpart of the 'state' in Delhi is the Government of the National Capital Territory of Delhi (GNCTD). It was constituted by the National Capital Territory of Delhi Act, 1991. The Act provides for a Legislative Assembly of 70 members that is directly elected by voters from as many constituencies. The inspiration behind the Act was the considered views of two committees appointed by the Government of India—the Balakrishnan Committee and the Sarkaria Committee, which held that Delhi's citizens should have the right to democratic participation in government, even as the city's Union territory status was maintained in the national interest.

The leader of the group that secures the majority in the election, the Chief Minister, chooses his ministers, who are then formally appointed by the President. Article 239AA of the Constitution of India, under which the Delhi Act 1991 was enacted, explicitly says that the Council of Ministers with the Chief Minister at the head will aid and advise the Lieutenant Governor in the exercise of his functions in relation to matters with respect to which the Legislative Assembly has powers to make laws. Subject to the constitution this excludes all matters on the state list and concurrent list except Entries 1, 2 and 18 on the state list and also Entries 64, 65 and 66 of that list in so far as they impact upon Entries 1, 2 and 18. Article 239AA also provides the Lieutenant Governor with discretion to refer any matter to the President of India for decision in the case of difference of opinion between him and his Ministers. Pending such decision, the Lieutenant Governor, can take any action or give any direction in a case where the matter, in his opinion, is so urgent that it is necessary for him to take immediate action.[29]

The dual responsibility of the Governor towards the State government on the one hand and the Central government on the other, together with the constitutional provision of discretionary authority to facilitate its discharge has been an abiding source of

---

29. Bakshi, P.M. (2000). *The Constitution of India*, Universal Law Publishing, p. 204-205 Delhi.

tension in India's federal system, particularly when different political parties are in power in the Centre and the States/Union Territories. This duality impacts upon Delhi government even more severely because else where the Governor does not have such an explicit discretionary power. In November 2002, during the fourth year of her first tenure, Sheila Dikshit, Delhi's Chief Minister, generally regarded as a very mature politician, went public with her complaint about the constant interference of the Lieutenant Governor in her domain and said that the Lieutenant Governor was stepping in at every opportunity and taking parallel meetings of departments and issuing policy directives contrary to hers.[30]

The executive authority of Delhi's Chief Minister is weak for another reason. While in the case of States their executive powers 'extend to the matter with respect to which the Legislature of the State has powers to make laws,' the Union Territories are 'administered by the President' through the administrator appointed by him. The NCT Act of 1991 does not provide for the transfer of executive powers to the Delhi government. The executive powers are in the hands of the President of India who can delegate them from time to time. At present, Marina Pinto observes the delegated powers are vested in "the Lieutenant Governor and in certain areas, with the Chief Secretary, who along with the Police Commissioner, the Home Secretary and Secretary, Land and Development, are appointees of the Central government. Hence, by law, the President continues to be the final authority in the realm of executive powers." Pintos' view is, "as per the NCT Act Delhi government hardly retains any powers over services. Since the services form part of the reserved section of the list which is within the ambit of the union, Delhi government cannot even appoint a peon leave alone the framing of service rules".[31]

Coming to legislative powers, since Delhi is the capital city and the Union Government has special responsibilities relating to security. This means, Delhi government cannot make laws on 'public order,' 'police' and 'land'. Right in or over land, land tenures including the relation of landlord and tenant, and the collection of rents; transfer and alienation of agricultural land; land improvement and agricultural

30. Vohra, Pankaj (2002). "What Will My Detractors Say Next?," *Hindustan Times*, November 24, Delhi Edition.

31. Pinto, M, *op. cit.*, p. 151.

loans; are implicitly excluded." These are substantial constraints and considerably limit the powers of the Chief Minister.

Besides what is more significant is that Delhi government, unlike other States, cannot frame its own rules for transaction of business. A former Chief Secretary of Delhi has stated that framing of the transaction of business rules is "the most important trappings of a State."[32] This lacuna enabled the Central government in 2002 to amend the Transaction of Business Rules, whereby the Delhi government could not introduce any bill in the Legislative Assembly without the prior approval of the Lieutenant Governor. On 25 July 2002, the Ministry of Home Affairs informed the Delhi government that all legislative proposals would have to be cleared by the Lieutenant Governor. Delhi government was further informed that, "it has been decided by the Central government that in the legislative proposals relating to the National Capital Territory of Delhi the word 'government' may be defined as the Lieutenant Governor of the National Capital Territory of Delhi." It was elaborated that there was no other government in Delhi other than the Central government and Article 239AA was a special provision for Delhi, which provided that the Union Territory would be called the 'National Capital Territory of Delhi' which did not bring a new and independent entity into existence.[33]

On August 29, a Presidential Order was issued, amending the Transaction of Business of the Government of the National Capital Territory of Delhi Rules, 1993, whereby the Lieutenant Governor's powers were enhanced to ensure that all classes of proposals or matters were submitted to the Lieutenant Governor. Though the Lieutenant Governor insisted in a press conference that there was neither any reduction in the powers of the Council of Ministers to make appointments nor any reduction in its financial powers and that the amendment would not apply to subordinate legislation, rules and guidelines, the Chief Minister considered it as a severe curtailment of the powers of the Delhi government because it meant that even before a Bill was introduced in the Assembly, it would have to be sent to the Lieutenant Governor for clearance, at the proposal stage.[34]

32. Sehgal, Omesh (2003). "Without Some More Powers, Delhi May Soon Reach the Point of No Redemption," *Hindustan Times*, January 24, Delhi Edition.

33. Rajalakshmi, T.K. (2002). "Capital Coup," October 12-25, *Frontline*, Madras.

34. *Ibid.*

Politics has determined how far the issue of Transaction of Business Rules is regarded as a constraint by the Delhi government on its legislative powers and a tool for the usurpation of its authority. In the early 1990s, when the Congress and the United Front coalition was in power at the Centre and the BJP controlled the NCT, the mere existence of the Rules were not basis for conflict. In 1998, when the BJP led coalition came to power at the Centre, it deleted it at the request of the then Chief Minister, Sahib Singh Verma, who was from the BJP. The Rules were reimposed in 2002 when the Congress was in power and elections were due in a years time. The timing made the decision controversial.

The Delhi government is not only limited in its functioning by the powers of the Central government, it is also limited in its control over local governments in its jurisdiction because of the Central government's powers over these bodies. The MCD and the NDMC are governed by Central government acts. The MCD, which was created by an Act of Parliament in 1957, has acquired the present status after the comprehensive changes that were introduced by the Delhi Municipal Corporation (Amendment) Act, 1993 in the wake of the passage of the Constitution (74th Amendment) Act of 1993. Many of the Central government's powers in the original Act have remained unaffected. The Commissioner of the MCD, its chief executive officer, is also appointed by the Central government. Though both the Central government and the Delhi government have the power to inspect any record, plan or document and any department and office of the MCD, the power of the Central government *vis-à-vis* the MCD is more real. In respect of Building Regulations, the Commissioner is required to exercise his powers and discharge his functions "under the general superintendence, direction and control of the Central government." The Central government can also dissolve the Corporation if it feels that it has "persistently defaulted in the performance of its duties or has abused its powers."[35]

Municipal Corporation of Delhi

The three local bodies in the National Capital Territory of Delhi are the Municipal Corporation of Delhi (MCD), the New Delhi Municipal Committee (NDMC) and the Delhi Cantonment Board.

35. Pinto, M, *op. cit.*, p. 140.

Among the three, the MCD is the largest local body with the responsibility of providing the basic amenities to both the rural and urban areas encompassing 1,399 square kilometres or about 94.22 per cent of the whole NCT area. The MCD claims that it is among the largest municipal bodies in the world providing civic services to more than an estimated population of 13.78 million citizens in the capital city and that it is next only to Tokyo in terms of area. The NDMC extends over an area of 42.74 square kilometres (2.88 per cent). This is where by and large the city's diplomatic and bureaucratic elite works and lives and accounts for 3.2 per cent of the population. The Delhi Cantonment Board covers an area of 42.97 square kilometres (2.9 per cent) and accounts for just 1 per cent of the population.

Unlike in the past when the local body could be arbitrarily dissolved or suspended, the Delhi Municipal Corporation (DMC) (Amendment) Act, 1993, provides for its almost continuous existence. The term of the Corporation is five years from the date of its first meeting. If dissolved earlier, the DMC (Amendment) Act provides that an election to constitute the Corporation must be completed before the term ends or within six months from the date of dissolution. The primary objective of the Constitution (74th Amendment) Act of 1992 was to introduce public participation and accountability at the local level. The DMC (Amendment) Act, 1993 therefore provides for division of Delhi in 12 zones for administrative purposes. Each zone has a number of wards. There are 134 wards in all. The population in the zone varies and there is no pattern in the numerical quantity of wards into which they are distributed. Shahdara (N) has 16 wards while Central and South zone have 12 each. In 1991, while 12.02 per cent in the MCD area lived in the first zone, 7.08 per cent and 11.77 per cent lived in the second and third zones respectively.

The present set up of the MCD consists of a legislative wing and an executive wing. The legislative wing has 134 elected members, who are elected on the basis of universal adult franchise, together with nominated members, who include Members elected from Delhi to the Lok Sabha, Rajya Sabha, the Legislative Assembly and people who have special knowledge and experience in municipal administration. For effective functioning of the Corporation, statutory bases have been provided for the Standing Committee, the Ward Committees that can be created for a municipality area that has three lakh or more residents, the Education Committee and Rural Areas Committee. The Standing Committee consists of six members elected by the Councillors from among themselves and one member elected by the members of each Ward Committee from among themselves. The Standing Committee looks after matters such as budget sanction of expenditure, approval of contracts and creation of posts.

Each zone has a Ward Committee that consists of all the Councillors elected from the wards in that zone. The 15th Schedule, which was added to the DMC (Amendment) Act of 1993 and which defines the powers and functions of the Ward Committees, suggests that a leadership role had been envisaged in the functioning. The Ward Committees have been designated as the municipal authority to deal with those civic matters that directly affect the people. Among other objectives, they sanction estimates and plans for municipal work to be carried out within the zone costing up to Rs. 1 crore and can call for any report, return, plan, estimate etc., concerned with matters pertaining to the zone. In spite of these vast powers, the Ward Committees have not proven to be very effective in building people's participation in municipal decision-making.[36]

The Mayor is the head of the Corporation and is elected by the elected members from amongst themselves every year. The Mayor presides over the deliberations of the Corporation and conducts its meetings.

The Commissioner who is appointed by the Central government for a term of five years is the head of the executive wing. He exercises all the powers and performs all the duties specifically conferred on him by the DMC (Amendment) Act of 1993 and prescribes the duties of all other municipal officers and controls and supervises their work. Six Additional Commissioners who are in-charge of various departments and zones assist him. The MCD has divided the city of Delhi into 12 zones for administration purposes and most departments have decentralised their operations to the zonal level. A Deputy Commissioner is the administrative head of each zone.

The DMC (Amendment) Act of 1993 distinguishes between the obligatory and discretionary functions of the MCD as per the format

---

36. Ghosh, Archana (2000). *Solid Waste Management in Delhi*, pp. 24–26, Institute of Social Science, New Delhi.

of the Constitution, 74th Amendment Act. It is incumbent on the MCD to make adequate provisions for the fulfilment of 22 obligatory functions. The discretionary functions have to be performed by the MCD if necessary resources are available. The obligatory functions include major aspects of solid and liquid waste management. These are:

- construction, maintenance and cleaning of drains, drainage works; public latrines, urinals and other public conveniences;
- scavenging, removal (which includes transportation) and disposal of filth, rubbish, other obnoxious and polluted matters;
- reclamation of unhealthy localities, removal of noxious vegetation and generally abatement of all nuisances; and
- lighting, watering and cleansing of public streets and other public places.

The Conservancy and Sanitation Engineering Department (CSE) of the MCD looks after the sanitation work in the areas within the jurisdiction of the MCD which includes urban villages, notified slum clusters, regularised unauthorised colonies, squatter settlements, streets and public conveniences. The sanitation work translates into sweeping an area of about 700 square kilometres, maintenance and removal of garbage from about 2,400 garbage dumps, cleaning/de-silting of sewers (approximate length 4,276 kilometres), except major sewers connected with Sewage Treatment Plants (STPs), cleaning/de-silting of drains (approximate length 1730 kilometres), operation and maintenance of 152 sullage pumping stations, excluding those connected with STPs, cleaning/de-silting of 578 major *nallahas* carrying sullage throughout the year, de-silting and disposal of liquid waste in 1,347 septic tanks mostly located in resettlement colonies and cleaning of public toilets and urinals.[37]

The Conservancy and Sanitation Engineering department (CSE) is placed under the charge of Additional Commissioner (Headquarters). Two Directors of the rank of Chief Engineer head the department. Director-1 is responsible for the CSE operations in 10 zones of the city west of Yamuna and Director-2 looks after the two trans-Yamuna areas to the east. Six Joint Directors of the rank of Superintending

37. *Ibid*, p. 27.

Engineer (SE), who are responsible for two zones each, assist the two Directors. In addition, there is one SE (Auto) and one SE (E and M) who look after the auto workshops and the electrical and mechanical maintenance operations respectively. An Executive Engineer heads the CSE operations at the zone level. Two Assistant Engineers (AEs) and two Sanitary Superintendents (SSs) assist each Executive Engineer (EE). The EEs, besides reporting to the Joint Director, also report directly to the Deputy Commissioners of the zones regarding field operations. In fact EEs, AEs and SSs, all have a dual responsibility and report to the Deputy Commissioners of the zone as well as to the Joint Directors of the CSE department.

There are about six to seven Junior Engineers in a zone who are mostly involved in the engineering aspects of maintenance of storm water drains and *nallahas* and repair and maintenance of *dhalaos* and dustbins as well as public toilets and urinals. The SS in the zone is responsible for sanitation operations which involve the removal of refuse and small dead animals and cleaning of open and covered drains, spraying of insecticide etc., at collection points and ensure transportation of the refuse to the landfills, assignment of tasks to drivers, requisitioning of vehicles and indenting of fuel. Under each SS there is usually one Chief Sanitary Inspector and six to eight Sanitary Inspectors (SIs) depending on number of wards in each zone. There are usually three to four Assistant Sanitary Inspectors (ASI) under each Sanitary Inspector (SI). Sanitary Guides (SG) assist the ASI. The ASI's and SGs supervise the work of the *Safai Karamcharis* (SKs).[38]

One of the reasons for the unsatisfactory discharge of civic functions, despite the existence of a dominant local government body with explicit legal powers and elaborate structure, is the confusion due to division of responsibilities between the MCD and related bodies which could be local, state or central. At the local level, there are two other bodies, the NDMC and the Cantonment Board (CB). At the next level, there is Delhi government and its departments such as the Flood Control Department (FCD) and autonomous bodies such as the Public Works Department (PWD) and the Delhi Jal Board (DJB). The Central

---

38. The Energy Research Institute (2002). "Urban Services Environmental Rating System" Final Report (Draft), Report No. 1999EE41.

government and its agencies such as the Ministry of Urban Development (MoUD), the DDA and the Central Public Works Department (CPWD) have jurisdiction in specific areas of Delhi. The DJB is responsible for providing trunk sewers and sewage treatment plants for whole of Delhi. In the areas of the NDMC, the Cantonment Board and the DDA development areas, these agencies are responsible for providing internal, peripheral and outfall sewers. Within the MCD area, trunk sewers and sewers connected to trunk lines are the responsibility of the Jal Board, the majority of the remaining sewers are maintained by the MCD and certain sewers are meant to be maintained by the Flood Control Department, which is under the Delhi government. Construction and maintenance of roads is under different agencies such as the PWD, CPWD, the DDA, the MCD, the NDMC, and the Cantonment Board. In the areas declared as development area, the DDA is responsible for building the infrastructure and providing the basic services. After the area is declared as developed, it is to be transferred to the MCD for maintenance. Often the 'developed' area has so many deficiencies at the time of transfer that it takes years for the formal transfer of the developed area to accepted by the MCD. During this period, maintenance work is neglected by both agencies. Lack of coordination between the various agencies has been the bane of Delhi.

Autonomous Bodies

Delhi Jal Board

Delhi Jal Board (DJB), an autonomous body of the Delhi government, was constituted through an Act of Delhi Legislative Assembly on 6th April 1998. It replaced the former Delhi Water Supply and Sewage Disposal Undertaking (DWSSDU), a body of the MCD created in 1958. Its management is constituted by ex-officio members and elected representatives. The CEO in recent years has been a senior I.A.S. officer. The Delhi Jal Board Act gives the DJB more powers and a larger autonomy than enjoyed by DWSSDU. It can for instance, call for private sector participation in its operations.

The DJB is responsible for the production and distribution of drinking water, treatment and disposal of sewage as well as water quality control. It provides water in bulk to the NDMC and the Delhi Cantonment Board for redistribution in their areas. Similarly the sewage is received in bulk by these two local authorities for final conveyance and disposal. In the areas under the jurisdiction of the MCD, the DJB is responsible for distribution of water and collection and disposal of sewage. Unauthorised colonies are not eligible for sewerage connection. Even in the remaining areas, compelling pressure rather than objective technical criteria determines which area is favoured.[39] One of the factors for the shortage of water and inadequate infrastructure for the upkeep of the sewage system is the low rate that is charged to metered household and non-metered households who make up 80 per cent of the consumers as compared to industry. Again electoral considerations inhibit a change in the *status quo*.[40]

Delhi State Industrial Development Corporation

Delhi State Industrial Development Corporation (DSIDC) was incorporated as a company and registered under the Companies Act in February 1971. Main activities of the Corporation are construction of industrial sheds, relocation of industries, construction of Common Effluent Treatment Plants (CETPs) for industrial estates and provision of marketing assistance to small-scale units. The commercial activities include liquor distribution, provision of technology consultancy, quality control laboratory, overseas manpower bureau and mining. The other objectives of DSIDC are to:

- Aid, counsel, assist, finance, protect and promote the interests of industries in the Union Territory of Delhi.
- Undertake procurement and distribution of various raw materials.
- Undertake and provide marketing facilities especially to the small-scale industries of Delhi.
- Acquire land and develop it suitably.
- Effect coordination between large industries and small industries.
- Promote and operate schemes for the industries of Delhi.

39. Sharma, Aruna P. (2002). "Lack of Coordination Plagues Bhajanpura," *Hindustan Times*, April 4, Delhi Edition.
40. Ruet, Joel, *et al.*, (2002). "Water and Sanitation Scenarios in Indian Metropolitan Cities-Resources and Management in Delhi, Calcutta, Chennai and Mumbai," *CSH Occasional Paper* No. 6, French Research Institute, p. 132, Centre de Sciences Humanies, New Delhi.

- Enter into arrangements with Government of India, Delhi government or any other government, or state, or local authority for the purpose of carrying out the objectives of the company.
- Promote and develop industries of all types including wholly Public Sector and Joint Sector Undertaking.
- Establish and maintain manpower data to export manpower to foreign countries.[41]

In 2002, DSIDC had a total strength of 1,094 persons. It earned a net profit of Rs 685.62 lakh in the year 2001-02. Until February 2005 three CETPs were constructed. DSIDC acquired 1,865 acre of land for relocating 26,000 units from residential and non-conforming areas. It has constructed about 840 industrial sheds for the purpose of employment generation since 1974. The Corporation runs 83 liquor retail outlets.

The Delhi Pollution Control Committee

In 1991, the Central Pollution Control Board (CPCB) delegated all the powers vested in it under the Water (Prevention and Control of Pollution) Act, 1974 and the Air (Prevention and Control of Pollution) Act, 1981, in respect of the Union Territory of Delhi to the Delhi Pollution Control Committee. At the time of its formation, the DPCC comprised of four members: Secretary, Department of Environment, Secretary, Local Self Government, Director (Environment) and a nominee of CPCB. After more than a decade, the DPCC was reconstituted in 2002 with 12 members to include five experts, for the first time from such disciplines as engineering, architecture and medicine.

Since its constitution in 1991, the role of the DPCC has enlarged considerably. It has also been entrusted with statutory duties under Water (Prevention and Control of Pollution) Cess Act, 1977 for assessment/levy/collection of Cess from local authorities and specified industries. Besides, DPCC has been vested with statutory powers, functions and duties under Section 5 of Environment Protection Act, 1986 as also under various Rules issued under Environment Protection Act, 1986 *viz*:

---

41. Delhi State Industrial Development Corporation, *www.dsidc.nic.in*

- Manufacture, Storage and Import of Hazardous Chemical Rules, 1989.
- Hazardous Waste (Management and Handling) Rules, 1989 and Amendment Rules, 2000.
- Chemical Accidents (Emergency Planning, Preparedness and Response) Rules, 1996.
- Biomedical Waste (Management and Handling) Rules, 1998 and Amendment Rules, 2000.
- Recycled Plastic Manufacture and Usage Rules, 1999.
- Noise Pollution (Regulation and Control) Rules, 2000.
- Municipal Waste (Management and Handling) Rules, 2000.

Until March 2005, though the sanctioned staff strength of the DPCC is 166, only 90 positions have been filled up. In the senior positions, out of 49 posts only 17 posts were filled. Out of the 41 posts of Senior Environment Engineers/Environment Engineers/Assistant Environment Engineers/Scientists, only 14 posts were filled up. At the lowest level of peons/security guards/lab attendants, out of the 18 sanctioned posts, 17 posts were duly filled.[42]

Central Government Organisations

Since police, law and order and land lie within the Centre's jurisdiction, Central government agencies such as the Delhi Police can take punitive actions against those persons or vehicles whose movements are deemed a public nuisance or a violation of traffic rules. The Union Ministry of Urban Development controls land in Delhi through its autonomous body, the Delhi Development Authority, created by the Delhi Development Act of 1957, which authorised this body to acquire, develop, and dispose of land. DDA can refuse or agree to release land for construction of waste related facilities such as landfills. Disciplinary action undertaken by the Delhi Police is generally regarded as high handed and arbitrary by waste workers in the unorganised sector.

The framework of Delhi's development has been Master Plans that are approved by the Parliament. The first Master Plan was formulated

---

42. Delhi Pollution Control Committee, *www.dpcc.delhigovt.net.in*

at the initiative of Jawaharlal Nehru who in 1955 assembled a team of the 'best available planners available anywhere in the world' for the purpose. The result was the First Master Plan (1960-1981), the first for a city in independent India. The DDA was created with the specific mandate for implementing the plan in which growth of Delhi was envisaged in a 'poly-nodal/poly-centric' form with the several towns around the capital also being developed. The Master Plan did not succeed in its objectives. The second updated version of the Master Plan and the perspective Development Plan for Delhi, 1981-2001, also did not unfold in the manner that was hoped.[43] The persistent problem of demand for regularisation of unauthorised colonies illustrates the difficulties. The Central government from the late-1960s onwards has announced the conditions on the basis of which these could be regularised. In 1994 the Delhi government recommended the regularisation of 1,073 colonies that had arisen prior to 31 March 1994. In 1998, the Delhi High Court directed the Central government to finalise the policy for regularisation, giving a definite cut-off date and the development charges payable for regularisation. Delhi government was unable to secure the regularisation of these 1,073 unauthorised colonies until December 2003 inspite of its persistent efforts. It was generally believed that the Central government stalled the move by stipulating the stringent condition that the colonies could not be regularised until the existing land rates and the penalty were paid. As a result, these unauthorised colonies remained deprived of basic municipal services.[44]

In 2003, a draft of the Master Plan 2001-2021, which called for regularisation of all unauthorised colonies and a high-rise growth model, was made available for public comment.[45] The Chief Minister's observation was that it was unrealistic and unsustainable from the point of view of the city's existing infrastructure. Parliament's approval is awaited for the new Master Plan.

In recent years, the Union Ministry of Environment and Forests has become the final decision-making authority in the matter of public

---

43. Mehra, K. Ajay (2003). "Governing India's First City" in Seminar on *Sustainable Development of Delhi*, June 16-17, p. 5, Institute for Human Development, New Delhi.
44. Vohra, Pankaj, *op. cit.*
45. Kanth, Pratyush and Shivani Singh (2003). "Good for Votes, Not for Delhi," *Times of India*, July 29, Delhi Edition.

access to common resources such as groundwater and river water as a result of the enactment of Environment (Protection) Act 1986 by which the Central government has assumed the overall responsibility of 'environment protection and improvement.' One of the first steps that the Central government took was the creation of a Central Ground Water Authority to check the problem of over-exploitation of ground water in the urban areas. However, the Central government is hamstrung by the fact that it is dependent upon the Deputy Commissioner in whose jurisdiction there is a violation of its orders for its enforcement. Further, it is not the only government agency that can regulate the ground water as per existing law. Delhi Jal Board can also do so as per the Delhi Jal Board Act of 1998.[46]

Future Trends

Widespread dissatisfaction with the way the local government was exercising its functions has convinced policymakers that there should be smaller municipal organisations with fewer functions. The Virender Prakash Committee appointed by the Congress government like the V.K. Malhotra Committee appointed by the BJP supported the notion of breaking up the present MCD into four or five units with alterations in their powers and jurisdiction to make them more focused in their functioning. The Virender Prakash Committee recommended that the smaller units of the MCD should confine themselves to purely local functions: all drains below 1,000 cusecs discharge and all public roads with the width of less than 30 meters should be maintained by the municipalities; municipalities should be responsible for removing encroachments on public land and demolition of unauthorised construction; all hospitals presently under the control of the MCD should be brought under the management of the Delhi government; and a Slum Improvement Board created to determine slum policy and upgradation of *jhuggis* and their relocation. The municipalities should implement the schemes proposed by the Board and provide civic facilities in these areas.[47]

However, from the point of view of the citizen interested in speedy corrective action for civic lapses, anxious about ready access to

---

46. Ghosh, Sutapa (2003). "Clueless in Capital," *Down to Earth*, p. 17, December 31, Delhi.
47. *The Times of India* (2001). "Points of Contention," June 17, Delhi Edition.

authority figures, these changes could only be of limited value. The segmentation of the MCD could further add to civic confusion by creating more authorities and more boundaries in the city. There could be resistance to acceptance of garbage from other zones by those that have landfill sites. What was most needed was a greater sense of commitment and transparency in decision-making by office holders. In this context the adoption of citizen charters or statements by 66 departments of Delhi government since 1998 through which people can know what to expect from the service provider in quantitative and qualitative terms was a step in the right direction. Another significant development has been the passage of Delhi government's Right to Information Act, 2001 by which grounds of decision-making can be accessed in large number of cases by payment of a small fee. Special progressive features of the Act are imposition of a fine on the official for delay in giving information and for giving wrong information as also provision of an independent appellate authority to whom appeal can be made if one is not satisfied with the information provided.

Structural changes in the MCD alone will not result in durable improvements in provision of basic civic services. The Delhi government needs to have more effective control over the MCD. In the existing arrangement, it seems as if at times the MCD considers Delhi government's initiatives in waste management as unwarranted interference in its constitutional responsibility and usurpation of its powers. The consequence is an attitude of non-cooperation that compounds the difficulties of citizens. The authority of Delhi government over its territorial jurisdiction needs to be strengthened. A former Chief Secretary's view is that the NCT Act, 1991 should be amended to enable Delhi government to frame its own Transaction of Business Rules and to provide for Delhi's inclusion in the List of States so that it can have its own civil service as a means for effective and complete control over State functionaries. The provisions that give the Lieutenant Governor the power to hold back the decisions of the Council of Ministers need to be removed. While law and order and land could remain with the Central government, Delhi's Legislative Assembly should have full powers to amend/frame laws on all other state subjects.[48] However, Delhi government seems to be of the view

48. Sehgal, Omesh, *op. cit.*

that this would not be enough—even control over land should be transferred to it.

Delhi—A Sustainable Development Perspective

What distinguishes Delhi from most other cities in Asia and Africa in the beginning of the 21st century is the fact that it has a well-considered perspective in place on the subject of its future development. Development, it was suggested, should be in a manner which is 'sustainable.' In the sense in which this concept has acquired currency since late 1980s, it means that 'economic growth' should be combined with 'social progress' and 'environmental protection'—these were the three legs of a tripod representing a model development paradigm. This perspective for Delhi which envisaged a 20 year growth plan is explained and elaborated in a document called "Delhi-21". This document was prepared as part of the Delhi Urban Environment and Infrastructure Improvement Project. The project was aimed at suggesting how Delhi could be made, not only more 'environmentally sustainable and liveable' but also an 'internationally competitive and productive city' This project was sponsored by the Government of the National Capital Territory of Delhi (GNCTD) and the Ministry of Environment (MoE), Government of India with Japanese funding through the World Bank. The Report was composed by a team of experts from various fields placed together by GHK International, UK and Operation Research Group, India. The team interacted with concerned officials, agencies, interest groups and individuals from civil society through seminars, workshops, discussions and presentations before finalising its report.

It was acknowledged that the achievements relating to the continuing pattern of economic growth had been impressive, but what was equally urgent, was the need to halt the inequality in access and deficiencies in social utilities and the trend of steady deterioration in the condition of the environment. In Delhi-21, words were not minced to describe this situation. It stated, "The environment, urban planning and management in Delhi will continue to suffer from the current disabilities, and will not cope with future prospects unless a number of fundamental and drastic changes are made in its governmental structure."[49]

---

49. *Delhi-21, op.cit.,* p. 85

The vision period of 20 years was chosen because according to the study an extended period was needed to make sustained urban and environmental improvements in an "incremental manner in line with local commitment, resource mobilisation and institutional reforms and capacity building—all of which take time to develop." There was a similar requirement for price and regulatory reforms. However, given the gravity of the situation in Delhi it was felt that "most of these policy and institutional reforms need to be accomplished in the immediate to medium term in the next five to ten years at the most."[50] Every opportunity within the existing system should be used to review policies and consider alternatives. "In fact, reforming the planning process itself, and not repeating the past mistakes or reinforcing them through current practices should initiate a reform process."[51]

The study noted that a large number of Delhi's residents lived in a state of material deprivation and that "12.4 per cent or 1.7 million persons have calorie deficiencies in their daily diet and 30 per cent or 4.2 million persons have virtually no formal infrastructure (i.e. they suffer from non-food poverty)." It was estimated that, "in total, approximately six million of Delhi citizens are living in poverty or deprived conditions of one form or another; 3.5 million in JJs and 2.5 million in service deficient areas." These slum residents were the major supporters of the urban economy through "their labour market contribution and informal production activities."[52]

The study further noted that in Delhi, "sickness from water and vector borne diseases recur annually" and also there was rising incidence of respiratory infections, cancer, heart problems, lead poisoning and other such conditions. What this meant was that the rate of pollutants in Delhi's air, water and water bodies was so enormous that natural biological and geo-chemical processes were unable to neutralise them. The study commented that though there was a considerable amount of information on the causes and consequences of environmental degradation in Delhi, a "strategic integrated and coordinated institutional approach is not in place to address these issues."[53] Social and environmental concerns needed to

50. *Ibid*, p. 94.
51. *Ibid*, p. 85.
52. *Ibid*, p. 9.
53. *Ibid*, p. 71.

be integrated into long term (10-20 years), medium term (5-10 years), and short term (0-5 years) perspectives on nature of infrastructure that would be required in the coming two decades.

The study's view was that for further development of Delhi the concept of National Capital Region (NCR) that had inspired planning endeavours in the 1980s should be revived. The National Capital Region Planning Board (NCRPB) should consult with GNCTD and the neighbouring state governments should distinguish between towns in Delhi Metropolitan Areas that may be regarded as satellite or dormitory towns and towns in the wider area, which will be more self-contained. The infrastructure requirements of the NCT area and the surrounding DMA towns for industrial estates, solid waste disposal sites, location of wholesale markets and transport terminals needed to be taken together. However, lessons needed to be derived from past experiences. NCR-NCT growth plan had little positive influence because the implementing agencies did not have clear strategies and the funding was inadequate. Further, "Physical planning should be regarded as a continuous process responding to the evolving needs of the community and the marketplace."[54]

Delhi-21 focussed on four types of urban infrastructure:

- housing,
- water,
- solid waste-municipal, biomedical waste, hazardous waste and by its emphasis on sewerage treatment, human waste, and
- transport.

Specific policies for sustainable growth were suggested in each of these areas. A common strategy was to bring about improvements by redefining roles of existing agencies and bringing in indigenous players whether they were CBOs or from the private sector. Thus, in the context of private housing, through a review of facts and figures, Delhi-21 trust showed that the DDA could not be considered as the major provider of formal housing. The housing policy needed to reflect the fact that the developers, builders and property owners in the private sector could build about six times faster than the housing

54. *Ibid*, pp. 21, 22.

institutions in the public sector. The housing policy was also required to devise means of reducing the number of vacant plots and facilitate entry of other players in the housing sector by encouraging the formation of Public/Private/Cooperative Sector partnerships to:

- Open up and service new areas.
- Upgrade slums and un-served settlements.
- Build capacity in finance mobilisation and mortgage insurance service.
- Review the impediments caused by the Rent Control Act and leasehold interests.
- Promote an open and transparent land and property market.

About *Jhuggi-Jhopri* Clusters (JJCs) and Un-served Settlements, the study acknowledged their existence as part of the urban scene in the Indian cities and recommended their upgradation and improvement as a matter of policy. According to the study the "growth and proliferation of JJCs can be reduced by providing a continuous supply of affordable serviced land, linked to public transport, and located near to industrial and other employment opportunities." About relocation, the study recommended, "relocation should be kept to the absolute minimum but some relocation may be necessary to provide land needed for civic amenity purposes" and "development of commercial activities."

The study also recommended a selective approach in "allowing expansion and locations of new offices, trade and institutional activities." Setting up of new offices and institutions and major expansion of exiting offices and institutions "that cause major addition to employment, traffic and transportation needs" were not to be allowed. The study recommended that only small offices which required high skill employment and trades which conformed to environmental standards should be located in Delhi.[55]

Structural Changes in Government Organisations

To remove civic deficiencies and mainstream environmental concerns, the first step was to strengthen the over-arching democratic political authority, *viz.* the Delhi government. "The duality of control

55. *Ibid*, pp. 26-28.

and separation of the lines of control and reporting have been creating problems of integrated and comprehensive sustainable development in Delhi." The study recommended that the MCD and the DDA should be brought under the administrative control of the GNCTD and the MCD split into smaller units. The splitting of the MCD into eight municipalities co-terminous with the revenue districts of the GNCTD "will set in motion the processes of (a) the convergence of the local administration between the municipalities and the revenue districts; (b) decentralisation and delegation of powers and functions; (c) providing a new basis for the applicability of the 74th Constitutional Amendment." The DDA should be split vertically and its planning functions "located in the GNCTD as a servicing unit for the Metropolitan Planning Committee as required in the 74th Constitutional Amendment." The development functions of the DDA needed to be "corporatised into a joint sector company with private sector participation to continue with the investment programme."[56]

Strengthening of Environmental Organisations

Framing and adopting an environmental policy for Delhi was the next step. A Cabinet Committee on Environment and Infrastructure for Delhi-21 headed by the Chief Minister and consisting of Ministers of Environment, Industry, Transport, Urban Development, Planning and Finance should be the nodal agency for resolving conflicts, monitoring overlaps and recommending authoritative steps for remedial action in the implementation of the policy. Since the work of the Department of Environment (DoE) and the DPCC in environment protection and improvement was dependent largely upon the activities of other government departments and agencies such as Industry, Transport, Urban Development, Health, PWD, DDA, MCD, NDMC, DJB, DTC, environment cells needed to be established in each concerned department and agency from within the existing staff members and steps taken to ensure that they become equipped with requisite skills.[57]

The study recommended that the DoE and the DPCC needed to be strengthened and their staff trained to "take on specific environmental management functions."

56. *Ibid*, pp. 85-86.
57. *Ibid*, p. 89.

The DoE was needed to be made responsible for policy formulation and implementation in areas such as "environmental planning and management; environment and industry; community participation; natural resource management; environment and transport (besides) public awareness" because these issues needed a higher level of coordination, not simply centralised control across all government departments.

The DPCC required similar attention if it was "to perform to a standard anywhere near that required for a major city such as Delhi." The DPCC needed to satisfactorily perform within its statutory remit in order to deliver, "regulation and enforcement; monitoring and data analysis; training outreach and support; (and) appeal procedures."[58]

Another key element in the action plan was the preparation by the DoE and the DPCC of a Citizens' Charter. This was "a statement from a public authority on what people can expect from the service provider in quantitative and qualitative terms." The Citizens' Charter could also provide people "an opportunity to interact with the public authorities, know the reasons for failure, when things might improve, and what improvements may take place in future."

The DoE and the DPCC it was stated, should also "provide a Public Participation Window for community-level local environmental management through CBOs and NGOs." This window should be utilised for interactions with the "trade unions, industry associations, professionals etc. wherever and whenever necessary."[59]

Direct private sector participation (PSP) and public-private partnership (PPP) offered a way to access new technologies and new sources of finance since private providers were more motivated to increase efficiency and raise profits from which benefits could be ensured to the consumer. However, the areas where and how the private sector could be efficient and effective "had to be pragmatically settled in a situation-specific manner."[60]

---

58. *Ibid*, p. 73.
59. *Ibid*, p. 89.
60. *Ibid*, p. 87.

Modification of Financial Control Systems

About resource mobilisation, the study recommended that the strategy should be "based upon the concepts of responsible financial management, value for money, and efficient resource pricing." Spending departments would have to increasingly rely upon themselves for generating funds to support their activities. Through the analysis of comparable situations, appropriate private public funding mix could be arrived at.[61] It was necessary to introduce "earmarked" funds or designated funds that were generated by enforcing development controls that could be ploughed back into improving that function e.g. levy of a "differential surcharge on sales tax on petroleum fuel graded in terms of their pollution" impact could be given to the DPCC.[62]

On certain kinds of activities that created a high risk for the environment, the monitoring and control agencies were justified in charging a higher rate directly linked to the higher cost of monitoring. According to the study, there was a strong case for establishing specific courts for rapid imposition of standard fines and to allow confiscation of assets of offenders.[63]

Under the polluter pays principle, "all polluters should have the prime obligation to take care of their own waste." This obligation could be first extended to commercial waste generators and subsequently applied to the residents by making them pay "the full costs of service." However, the application of the principle of the 'user pay charges' on the residents would have to be nuanced so that there was a minimal impact on those who could least afford to pay.

The financial base for radical infrastructure improvement programmes in Delhi was more than adequate. In comparison with other Indian cities Delhi could be said to be a highly productive and a rich city despite the presence of urban poverty. The economy had benefited from the process of economic liberalisation. The GNCTD's main tax base, sales tax, was buoyant reflecting the city's economy. Though the Delhi government had earmarked 9.5 per cent of its taxes for sharing with its local bodies to augment their resources, the local

61. *Ibid*, p. 75.

62. *Ibid*, p. 93.

63. *Ibid*, p. 75.

bodies—MCD, NDMC, DJB—were financially weak primarily because of distortions in the pricing of their services. The local bodies needed to tap their financial potential. In the case of MCD, the need for property tax revision was well recognised and was indeed long overdue. Through institutional reforms such as a voluntary retirement scheme, establishment costs of local bodies could be reduced. The local bodies needed to adopt a more commercial orientation in their functioning and levies for the services provided on basis of full costs. The traditional cash based accounting system needed to be modernised to introduce more transparency and the budget planning needed to more accurately "reflect the strategy, priorities and service outputs." The water cess rates that are set nationally were last fixed in 1992 at very low rates. A change in legislation was required to enable the States and the Union Territories to fix their own cess rates so that cities such as Delhi could fix cess rates related to the prevailing conditions. A categorical assertion made in the study was that "for environmental improvements borrowings could be contemplated only if there were parallel action programmes of policy and management leading to financial strengthening."[64]

The study acknowledged that fundamental changes required were political in nature and "their acceptability and implementation will take some time through discussions and processing." However, it suggested that no opportunity should be lost in carrying out a review of the existing arrangements for the purpose of identifying and undertaking comprehensive policy and structural changes since quality of city life and environment could not be improved through a sectoral emphasis. "Government, industry and communities must develop and achieve a shared vision for Delhi through an integrated environment regime. This will operate to produce a win-win situation where the four strands of sustainable development (economic development, social progress, environmental protection and resource minimisation) are in harmony. It will produce a climate where Delhi takes its place in the international community as a bankable, competitive and liveable city."[65]

---

64. *Ibid*, pp. 92-93.
65. *Ibid*, p. 71.

2

From Solid Waste Management to Sustainable Solid Waste Management

> "The streets themselves are usually unpaved and full of holes. They are filthy and strewn with animal and vegetable refuse. Since they have neither gutters or drains the refuse accumulates in stagnant stinking puddles." – *Frederick Engels, The Condition of the Working Class in England.*

Visitors from industrialised countries to low-income countries since the early 1980s until the end of the last century have found it convenient to quote the same words written in 1844 about 19th century London to illustrate their understanding of conditions in large parts of big cities in Asia and Africa.[1] The sub-text clearly is that while countries in what was till recently regarded as the 'Third World' were quick to follow in the footsteps of the developed world in industrialising their production processes and creating supportive human settlements, they were, in fact, proving to be increasingly less capable of managing the consequences of their initial decision to become a part of the modern world. A minimum standard of cleanliness in public spaces was not an unreasonable demand.

Insights into the constraints these countries faced and continue to face are not lacking. Urbanisation levels in less developed countries between 1975 and 2000, are considered similar to those of Europe, North America and other more developed regions between 1900 and 1925 when rate of urbanisation increased from 26 per cent to 40 per cent. Yet there are at least four ways, analysts say, in which the urbanisation process is different in most less developed countries from

---

1. Thomson, A.M. (1984). "Urban Poverty and Basic Needs: The Role of the Public Sector," in Richards, P.J. and Thomson, A.M. (eds.), *Basic Needs and the Urban Poor: The Provision of Command Services*, Crown Hamilton, p. 9 and Chaplin, Susan E., (1999), "Cities, Sewers and Poverty: India's Politics of Sanitation," in *Environment and Urbanisation*, Vol. 11, No. 1, p. 145.

that which occurred in Europe and the United States of America. Urbanisation is taking place at lower levels of economic development; it is more vulnerable to changes in international economy; it is based on a lower mortality and a higher fertility rate; and because of the higher migration from the hinterland, it involves many more people.[2]

Further, while in developed countries, the number of cities with at least one million inhabitants increased moderately between 1975 and 1995 from 85 to 114, in the same period, the number of million plus cities in less developed countries soared from 110 to 250.[3] Against this background, there could be only limited funds and skilled human resources left for provision of basic infrastructure when economic growth process itself was both technologically complex and capital intensive. With competing demands from sectors such as housing, transportation and energy, it was understandable that the management of unwanted material haphazardly abandoned within municipal limits should have low priority.

In reality low domestic priority did not necessarily mean a low level of means to take care of the waste. In mid-1950s, in an ideologically divided world, this situation drew the sympathetic attention of the advanced industrialised countries. Automatic loading trucks, bulldozers, incinerators and technology for centralised sewerage system was made available to the developing countries even as there was increased and focussed expenditure by developing countries on collection, transportation, treatment and disposal of waste, or 'solid waste management.' Transition towards a multi-polar globalised world did not stem the flow of friendly assistance because the industrialised countries by then had begun to panic about 'dirty polluting' ways in the poorer countries impacting upon them through the medium of an apparently seamless environment. However, there were no remarkable improvements in the developing world. As recently as in 2001, it was observed that, "despite consuming a significant share of municipal budgets (often between 10 per cent and 50 per cent) of operational expenditures, solid waste management services in the towns and cities of most low and middle countries are unreliable, provide inadequate

---

2. Brockerhoff, Martin P. (2000). "An Urbanising World," *Population Bulletin*, Vol. 55, No. 3, *www.prb.org*

3. *Ibid.*

coverage, conflict with other urban services and have adverse effects on public health and the environment."[4]

In retrospect it seems that solid waste in a low-income country is in a class by itself. It has become an intractable problem not so much due to a lack of technology or funds for disposing it safely but the manner in which it has been conceptualised and the choice of strategies for dealing with it.

Along with the model of fossil fuel based industrial production and consumption from the industrialised countries, the developing world also imbibed from it, its understanding of waste or residual objects which are a consequence of this process. It also accepted, by and large, the strategy of industrialised countries for handling and treating and disposing waste. The nature and composition of solid waste was idealised from the historical situation in early decades of the 20th century Europe when food processing industry developed to meet the needs of packaged ready-to-eat meals for the families of the factory workers and waste came to mean a non-hazardous mixture of paper, plastic, glass, metal and food residues.[5] Though the first refuse incinerator had already come into being in 1878, the equipment that solid waste came to be closely identified with, such as compaction trucks and automatic load vehicles was developed only in the first half of the twentieth century.[6]

In low-income countries, solid waste is much more heterogeneous than in the West. In these countries, besides containing a larger percentage of organic material, it has pathogen laden human excrement and toxic chemicals as its components. Human waste is absent in the waste aggregate in the industrialised countries because of the expansion of water-borne sewerage system from the mid-nineteenth century onwards. For hazardous waste, the industrialised countries have had the means to formulate comprehensive rules and ensure compliance when dangers from careless disposal of specific substances became evident from the early 1960s. Unlike in the

4. McDougall, Forbes, *et al.* (2001). *Integrated Solid Waste Management—A Cycle Inventory*, p. 27, Blackwell Sc Limited, Oxford.

5. Winiwarter, Verena (2002). "History of Waste" in Bisson, Katy and Proops, John (eds.), *Waste in Ecological Economics*, p. 50, Edward Elgar.

6. Barata, Eduardo, (2002), "Municipal Waste", in Bisson, Katy and John Proops (eds.), *op. cit.*, p. 119.

industrialised countries, the countries in the developing world had indigenous informal networks that routinely engaged in collecting, reusing, recycling waste which was capable of reducing the municipality's load. But public policy, by and large, considered it impractical to incorporate these in the official waste management system. Similarly, the limited reach of the capital intensive sewer based system and the unaffordable costs of maintaining huge mechanised vehicles which could not negotiate narrow streets for the purpose of garbage collection did not deter developing countries from investing even further in these. It was no surprise therefore, that the results of their labour at improving existing arrangements were dismal, both from the point of view of protecting public health and reducing dangers to the environment, the two well established objectives of a solid waste management system.

At Rio de Janeiro in June 1992, the industrialised world presented the developing countries with a vision of a 'sustainable' waste management system, one which was in line with the physio-chemical character of their waste, their socio-economic realities and technological capacities and at the same time sensitive to their aspiration for higher levels of economic growth and mass consumption. Detailed action plan was also put forward together with a generous offer of financial assistance for its implementation.

The money never came. An interesting development, however, since the mid-1990s was the endorsements of these principles in developing countries not merely by committed individuals and non-governmental organisations but also by government funded research institutions and politically influential teams of experts such as those appointed by the Supreme Court and the Planning Commission in the case of India. The result was that officials in key positions in waste management departments incorporated these principles in their policies and decisions. The process haltingly and unevenly spread across to various categories of solid waste. A new firm and encouraging trend in governmental decisions related to waste was clearly evident by the beginning of the present century in large parts of the developing world.

Understanding Solid Waste

Waste has been described as "useless remains." (Concise Oxford Dictionary). Waste is a by-product of human activity. It is material that is abandoned or discarded because the person who has it or owns it or is in a position to extract value from it chooses to rid himself of it. Waste that is neither atmospheric emissions nor a wastewater discharge is regarded as solid waste. In form this can be not only solid, but also semi-solid and liquid. In an urban setting, waste has been described as "material for which the primary generator or user abandoning the material within the urban area requires no compensation upon abandonment."[7] Urban solid waste is called municipal solid waste (MSW) because the municipality, the governmental authority at the local level due to its proximity to the site of waste generation, is considered best placed to collect it for the purposes of disposal.

Solid waste is an urban territorial phenomenon. Economic history shows that it emerged when human concentrations became engaged in non-agricultural forms of production. In subsistence agriculture based economies, residues were considered potentially recyclable and therefore a resource. In the earliest urban societies the form of waste that posed a major challenge to administrative authorities was that of human excrement.[8] It is interesting to note that in pre-modern times, an Asian city was able to solve this problem in a simple low cost manner, while a European city which was similarly situated could do so only through an elaborate material intensive high cost arrangement. In Edo, which later grew into the mega metropolis Tokyo, and which stood on several streams, all human waste was collected and returned to the farmland. The river remained clean for the water supply. Rome, which discharged its human waste into the Tiber, built long aqueducts to bring drinking water from distant sources.[9]

The criteria of an urban area vary from country to country.[10] The official definition of 'urban' in India is any place with a minimum population of 5,000; at least 75 per cent of the male working population involved in non-agricultural activities, and a population density of at least 400 per square kilometre. In practice, an area could

---

7. Cointreau-Levine, Sandra (1982). *Environmental Management of Urban Solid Wastes in Developing Countries*, p. 4, A Project Guide for the World Bank staff.
8. Winiwarter, Verena, *op. cit.*
9. *Down to Earth* (2002). "Sewer Systems Totally Destroy Nature's Nutrient Cycle," February 28, New Delhi.
10. Hardoy, E. Jorge, Diana Mitlin and David Satterthwaite (2001). *Environmental Problems in an Urbanising World*, p. 28, Earthscan Publications Ltd., London and Sterling, VA.

have all these characteristics and still not be considered urban because of deliberate retention of the rural status since such a categorisation entitles residents to benefits such as lower property tax and subsidised supply of water and electricity.[11]

Sandra Cointreau, who prepared a project guide for the World Bank on Environmental Management of Urban Solid Waste in Developing Countries in 1982, identified sources of solid waste as, households, commercial and industrial premises; institutions, and public streets. It was noted that street refuse contains a mixture of refuse from many sources as streets are used as dumping grounds by all and sundry.[12]

Sandra Cointreau's list of sources of solid waste is similar to that an analyst in a low-income country drew up almost 20 years later. The second person included rejected construction material and slaughter house waste as additional sources of wastes in towns and cities in India.[13] In the case of the United States, one scholar's list of municipal solid waste sources was confined to "small and moderately sized solid waste items from homes, businesses and institutions" that are "picked up by general collection trucks, typically compactor trucks, on regular routes" and to "bulky waste" consisting of larger items of solid waste such as mattresses, appliances, as well as smaller items generated in large quantity in a short time, such as roofing shingles or construction and demolition (C&D) waste, whose deposition at the municipal facility had to be especially arranged for.[14] Among recent studies of solid waste sources in Organisation of Economic Cooperation and Development (OECD) countries, in a European scholar's list of sources, even construction and demolition waste was not included in MSW. It was confined to only household and commercial waste.[15]

Solid Waste in Low-income Countries

The comparison of sources of waste in a low and high income country may suggest that since the list of sources is longer in the case

11. Beukering, Pieter Van, *et al.*, (1999). "Urban Solid Waste in Developing Countries: A Perspective on Bangalore, India," p. 9, *Working Paper No. 24*.
12. Cointreau-Levine, Sandra, *op. cit.*
13. Palnitkar, Sneha, "Municipal Solid Wastes," in Palnitkar, Sneha *(ed.) Manual on Solid Waste Management*, p. 10, All India Institute of Local Self Government, Mumbai.
14. Rugg, F. Mack (2000). "Solid Waste," in David H.F. Liu and Bela G. Liptak (eds.), *Hazardous Waste and Solid Waste*, Lewis Publishers, p. 162.
15. McDougall, Forbes, p. 2.

of a low income country more solid waste is produced in a developing country as per unit of population. Since waste is a by-product of production and consumption, and the scale of these processes is greater in a high income developed country the opposite is the case. Due to the nature of the waste management arrangements, in a developed country municipal solid waste accounts for only a small percentage of the total solid waste stream. In 1997, in the countries of the OECD, it was less than 10 per cent of the 522 million tonnes generated from sources which apart from MSW included agricultural and mining wastes, quarrying wastes, waste from energy production, waste from water purification, construction and demolition waste and manufacturing and industrial wastes.[16] In a developing country the proportion of MSW is considerably higher because safe and scientific disposal arrangements for various types of solid waste are not in place and waste from multiple sources is abandoned in municipal limits.

In a developed country, management of solid waste poses an exciting political and technological challenge. As a waste category that the general public (and therefore voters) have contact with the function of management of MSW has a high political profile. Additionally, household waste is, by its nature one of the hardest sources of waste to manage effectively. It consists of a diverse range of materials (glass, metals, paper, plastic organics) totally mixed together. MSW composition is also variable, both seasonally and geographically. It varies from country to country, and within countries from urban to rural areas. In contrast, commercial, industrial and other solid wastes tend to be more homogeneous and involve larger quantities of each material. Thus, if a system can be devised to deal effectively with the materials of household waste, it can be considered possible to apply such lessons to the management of other sources of solid waste as well.[17] In a developing country, the political benefit of sustainable improvements in waste management are overshadowed by those offered by competing options which merely make waste invisible such as removal of squatter settlements on public spaces or construction of a high wall around a community park, steps which ironically could interfere with a design for ensuring more durable improvements. The

16. *Ibid*, p. 2.
17. *Ibid*, p. 3.

technological aspirations of waste managers are likely to revolve around acquiring more sophisticated high visiblity equipment which may not even be appropriate for managing waste in a relatively backward setting. In recent years, not only has waste in developing countries steadily grown in quantity, it has increasingly acquired a hazardous dimension, since developing countries have become dumping grounds for electronic goods that have become obsolete in the industrialised countries. According to Exporting Harm: The High Tech Trashing of Asia, Report of the Basil Action Network (published in 2002), traders in e-Waste consider India, China and Pakistan among their favourite destinations.

The nature of waste in a low income has been consistently different from that of a high income country over the last two decades, in two respects, as evident from studies carried out on the subject of waste composition with in a gap of 17 years. It is organic based, more moisture laden and dense in comparison with waste in industrialised countries where the paper and plastic content is higher. Secondly, it has a component of hazardous pathogen laden material due to dearth of adequate sanitation arrangements. In recent years with expansion of indigenous industry in developing countries, a third difference, with respect to the presence of hazardous chemical waste has become evident.

A comparison of studies about solid waste in low and high income countries carried out in 1982 and that of four cities in low income countries—Karachi, Faisalabad, Addis Ababa and Arba Minch—in the late 1990s show that solid waste in cities of low income countries has continued to differ consistently and significantly from the refuse of industrialised countries. Table 2.1 shows ranges of various municipal waste components and 'moisture' and 'density' characteristics of such waste for cities in low income and industrialised countries in 1982.[18] Table 2.2 shows ranges of select municipal waste components and 'density' and 'moisture' characteristics of such waste in four cities in low income countries and cities of industrialised countries in the late-1990s.[19]

It can be noted that there is virtually no change in the range of moisture content and density in the waste of low income countries

---

18. Cointreau-Levine, Sandra, *op. cit.*, p. xii.

19. Ali, Mansoor, *et al.* (1999). *Down to Earth*, Loughborough University, UK, p. 14.

Table 2.1

Solid Waste in Low-Income and Industrialised Countries

Sl. No.	*Characteristics/Composition*	*Low-Income Countries**	*Industrialised Countries*
1	Waste generation (kg./cap/day)	0.4 to 0.6	0.7 to 1.8
2	Waste densities (wet weight basis-kg/cubic meter)	250 to 500	100 to 170
3	Moisture content (% wet weight at point of generation)	40 to 80	20 to 30
4	Composition (% by wet weight)		
	i Paper	1 to 10	15 to 40
	ii Glass and ceramics	1 to 10	4 to 10
	iii Metals	1 to 5	3 to 13
	iv Plastics	1 to 5	2 to 10
	v Leather and rubber	1 to 5	—
	vi Wood, bones and straw	1 to 5	—
	vii Textiles	1 to 5	2 to 10
	viii Vegetable/putrescible	40 to 85	20 to 50
	ix Inerts	1 to 40	10 to 85

Note: * Low-income countries are classified on the basis of their Gross National Income (GNI) referred to as Gross National Product before 2001 by World Bank that varies from time to time. In 1978 it included countries having a per capita income of less than USD 360. Since 2001, it includes countries with GNI of USD 900 or less.

Table 2.2

Solid Waste in Low-Income and Industrialised Countries: Select

Sl. No.	*Characteristics/Composition*	*Low-Income Countries**	*Industrialised Countries#*
1	Waste densities (kg./cubic meter)	300 to 400	150
2	Moisture content (%)	40 to 80	20 to 30
3	Composition (%)		
	i Paper	1 to 4	20 to 50
	ii Plastics	1 to 6	5 to 10
	iii Ash and fines	17 to 62	3 to 10

Note: * Case study cities

Data reported in Bhide and Sundsaresan, 1984.

between 1982 and the late 1990s. This is because the composition of municipal waste in the low-income countries has continued to be largely organic with the portion of vegetable/putrescible materials having a high moisture content. The relatively high percentage of non-putrescibles, paper and plastic, metal and glass that have large void spaces ensure a low density in a high-income country's solid waste. The degree of inerts in low income country waste contributes to its high density. Though paper and plastic consumption has increased in both rich and poor countries since the 1980s, there was only a marginal rise in the range of plastic waste in the municipal waste in low income countries. It was 1 to 5 in 1982 and 1 to 6 in late-1990s in the four cities. In the case of paper there was a decline, from 1 to 10 in 1982 to 1 to 4 in the late-1990s. This was because waste-pickers generally retrieve paper and plastic from garbage dumps for recycling. Thus, due to the intervention of rag-pickers, solid waste in a low income country has a distinct bio-degradable profile.

About street sweepings, Cointreau had observed in 1982 that this, "category of waste always includes dirt and litter. However, in developing countries it may also contain appreciable amount of household refuse, drain cleanings, human faecal matter and animal manure.... In any city where the sanitation services are inadequate, a portion of the population will directly employ open drains and roadsides for release of faecal matter."[20] In 2001, in a study about the environment problems of cities in Africa, Asia and Latin America, it was noted that about two-thirds of the urban poor in these countries and the Caribbean have no hygienic means of disposing of excreta. "Households with no provision for individual or shared sanitation within their homes have only three possibilities—public toilets, defecation outside, or defecation into some container which is then thrown away (what in some cities is called a 'wrap and throw' or 'flying toilets').[21]

Another difference between the MSW in high and low income countries, which has acquired prominence in recent years, is the presence of hazardous chemical residues from factories that is co-disposed with municipal waste since regulatory and enforcement systems in low income countries may be non-existent or non-

20. Cointreau-Levine, Sandra, *op. cit.*, p. 5.

21. Hardoy, Jorge E, *et al.*, *op.cit.* , p. 58.

operative. E-waste, electronic goods like computers, air conditioners, television sets, refrigerators, stereos and cellular phones in the industrialised countries that are regarded as dysfunctional because of mechanical failure or new technology advances have added a new dimension to hazardous waste in low income countries where they are in great demand due to a flourishing market in recycled goods. Companies in developed countries constrained by disposal codes inspired by the Extended Producer Responsibility principal are glad to make these available cheaply. E-waste contains toxic substances such as lead and cadmium in circuit boards; lead oxide and cadmium in monitor cathode ray tubes (CRTs); mercury in switches and flat screen monitors; cadmium in computer batteries; polychlorinated biphenyls (PCBs) in older capacitors and transformers; and brominated flame retardants on printed circuit boards, plastic casings, cables and polyvinyl chloride (PVC) cable insulation that release highly toxic dioxins and furans when burned to retrieve copper from the wires.[22]

Solid Waste Management in Low Income Countries

The consistency in the unique character of waste, which continues into the present, assumes significance in the context of designing arrangements for managing solid waste. Solid waste management (SWM) involves collection, transportation, treatment and final disposal of waste, which is regarded as being solid by the municipality.

Since the early-1980s till the present, SWM is regarded as a service to the public. In 1982 it was being performed quite unsatisfactorily. Sandra Cointreau in her study on urban solid waste in developing countries noted that it was, "frequently a source of complaints by the public and anxiety to the concerned officials."[23] In March 1999, the Report of the Committee appointed by the Supreme Court of India under the chairmanship of Asim Barman to suggest improvements in the solid waste management practices in the country noted that "solid waste management is an obligatory function of Urban Local Bodies in India. However, the service is poorly performed, resulting in problems of health, sanitation, and environmental degradation. With over 3.6 per cent annual growth in urban population and the rapid pace of

22. Puckett, Jim, *et al.*, (2002), *Exporting Harm: The High Tech Trashing of Asia*, The Basel Action Network.

23. Cointreau-Levine, Sandra, *op. cit.*, p. x.

urbanisation, the situation is becoming more and more critical with the passage of time."[24] Cointreau's view was that though provisions for collection and disposal services for municipal refuse consumed as much as 20 per cent to 40 per cent of municipal revenues, 30 per cent to 50 per cent of the solid waste generated was often uncollected.[25] In India, a 1990 study estimated that while between 10 per cent to 40 per cent of the municipal budget was utilised for SWM, on an average as much as 30 per cent of disposed solid waste was left uncollected.[26] van de Klundert and Anschutz on the basis of their studies on status of urban waste in low and middle income countries between 1995 and 2001 were of the view that, "waste collection and street cleaning often consume 10 per cent to 20 per cent of the total budget of a municipality in a year. Inspite of this, on average up to 50 per cent of urban dwellers in cities in low and middle income countries have no regular collection source (on average around 50 per cent of the population and even less in low income areas)."[27]

An implicit norm in the evaluation of SWM in low-income countries is the standard achieved in industrialised high income countries. In most industrialised countries in the matter of collection and disposal, services have expanded to the extent that over 90 per cent of the population (and 100 per cent of the urban population) have access to waste collection.[28] As for final disposal, about 70 per cent of the solid waste including hazardous waste is disposed of by burial in controlled landfills.[29] In such landfills, operations are so handled that short term and long term environment risks are reduced to an acceptable level.[30] Incineration and recycling also have a key role in reducing solid waste and thereby making it more manageable particularly in densely populated countries such as Japan and Netherlands. Incineration, however, is considered a very expensive option because apart from the cost involved in recruiting highly

---

24. Report of the Committee constituted by the Hon. Supreme Court of India (March 1999), *Solid Waste Management in Class 1 Cities in India*, p. 5.
25. Cointreau-Levine, Sandra, *op. cit.*, p. x.
26. Beukering, Pieter Van, *op. cit.*, p. 10.
27. Arnold van de Klundert and Justine Anschutz (2002). *Integrated Sustainable Waste Management—The Concept*, p. 23.
28. Beukering, Pieter Van, *op. cit.*, p. 7.
29. *Ibid*, p. 7.
30. Barata, Eduardo, *op. cit.*, p. 138.

skilled operators and careful maintenance that the equipment requires, mandatory initial expenditure to cover pollution abatement systems could be one third or more of the total investment.[31]

Recycling can be an impractical and irrelevant option in the industrialised countries because the financial and energy cost of collecting, transporting, segregating and reprocessing the waste to retrieve the natural material for new use could be more than the purchase and utilisation cost of virgin material.[32] In recent years the advance of less material intensive products and eco-efficient technologies in the industrialised countries has reduced the importance of the recycling option.[33] In contrast, in developing countries the prevalent methods of solid waste disposal is through uncontrolled dumping or burning on open ground or city streets.[34] In this process valuable resources that can and should be retrieved get lost.

In a low income country though there could be regular collection and removal and some reduction through composting, benefits to public health and environment are likely to be limited due to neglect of arrangements for final disposal at landfill sites. Even where landfills exist, the disposal of solid waste at such sites is not done properly. About India where many large cities have designated sites with controlled entry to serve as landfills, the Barman Committee was of the view that, "disposal of waste is done in a most unscientific manner. Generally crude open dumping is adopted for disposal of wastes in a low-lying area. Most local bodies deposit waste at the dump yard without ascertaining the suitability of the land for waste disposal and do not bother to cover the waste with inert material."[35] Capital intensive engineered structures that are required to minimise the impact of leachate on ground water and of the explosions due to methane generation are invariably missing. The fact is that there has to be a reliable revenue stream to ensure maintenance of the solid waste disposal system and that in turn has to be linked with an effective primary (e.g. from the waste generator) and secondary (e.g.

---

31. *Ibid*, p. 133.
32. *Ibid*, p. 120.
33. McDougall, Forbes, *op. cit.*, p. 4.
34. Beukering, *op. cit.*, p. 7.
35. Report of the Committee constituted by the Hon. Supreme Court of India, *op. cit.*, p. 6.

from the waste collection centre, such as the *'dhalaos'* in Delhi. In high income countries, while ensuring health and protection of environment have become matters of routine, in low income countries these are neglected issues. Garbage heaps are breeding grounds for flies, mosquitoes and rodents. Smoke and stench and landfill gases, particularly methane, pollute the air and leachate generation threatens ground water supplies.

Waste Management Hierarchy

One of the weakest links in the recycling chain in low income countries in comparison with the high income countries relates to the neglect of recycling of post-consumer waste. There is a historical reason for this. In the early 1970s and after the first oil crisis in the wake of Arab-Israel war, when the industrialised world became generally alarmed about the finite character of natural energy sources, to which it could have ready access, decision-makers used a waste management hierarchy as a tool for identifying and pursuing strategies for various components of waste so that only a minimum was left for final safe disposal. There was general agreement that this hierarchy involved:

- Prevention or reduction of waste.
- Reusing waste in its existing form.
- Processing waste so that it could be used as a direct or indirect input for new products and its biological treatment.
- Incineration, ideally with energy recovery.
- Land filling so that there were minimal adverse environmental impacts.

In low income countries, like in the rest of the developing world, the view in the early-1970s was that an official policy to regulate waste was likely to create confusion and divert national energy from the goal of industrialisation and facilitation of mass scale production of consumer goods. Care for environment was considered to be in conflict with development goals and poverty alleviation. Mrs. Indira Gandhi, the then Prime Minister of India, summed up this sentiment in 1972 in Stockholm at the United Nations Conference on the Human Environment when she suggested that in countries like hers waste management was a non-issue. "Are not poverty and need the greatest polluters? How can we speak to those who live in villages and in slums about keeping the oceans, the rivers and the air clean when their own lives are contaminated at source? The environment cannot be improved in conditions of poverty. Nor can poverty be eradicated without the use of science and technology."[36]

A legacy of this view is that though low income countries are resource poor, there is no official policy to recover raw materials from waste. In highly industrialised countries, recycling activities in recent years have witnessed an increase due to high costs of making deposits in engineered landfills due to increasingly stringent standards related to emissions from incineration and opposition to proximity of landfills inspired by the Not In My Backyard (NIMBY) syndrome. In low income countries, recycling is carried out in the informal sector. It is demand and market driven and it does not optimise the recovery of materials. It, however, relieves the municipality of the obligation of collecting, transporting and disposing of a substantial quantity of waste.

Through the 1980s and 1990s, the challenge to municipalities posed by the task of collecting, transporting and disposing solid waste so as to minimise the hazards it posed to human health and the environment became more and more daunting. With increase in urbanisation levels and GDP growth levels there was not only a quantitative increase of the waste but its physical-chemical nature also became more complex. The percentage of packaging related waste—paper, glass and metal—went up. However, since the housing and sanitation arrangements were not keeping pace with the increasing urbanisation, it meant that except in exclusively elite areas, this apart from being mixed with the biodegradable kitchen waste was likely to become contaminated with human excrement. Its hazardous character was compounded due to the proliferation of health care facilities and the co-disposal of biomedical waste with municipal waste. Finally, the virtual absence of or weak enforcement of regulatory control on small scale units manufacturing petroleum products, paper, electrical goods among other objects meant that toxic chemical waste was also being dumped with household waste.

36. Nadkarni, M.V. (2002). "Environmental Concerns and Management in the South and the North—A Comparative Perspective," in (ed) S.N. Chary and Vinod Vyaslu (eds), *Environmental Management*, p. 282.

Towards an Alternate Solid Waste Perspective

Since the late 1980s, inadequacies of waste management services by the local government and the consequent neglect of public health and environment considerations prompted four different entities to take the initiative to improve the conditions. These were non-governmental organisations and community based organisations; private sector; central government and international agencies. Though there were some changes for the better, without an alternate holistic perspective which could convincingly show the need for adoption of an identical approach for the major constituent groups of waste, these gains could only be haphazard and transient.

Role of NGOs/CBOs

Right across Asia and South America, non-governmental and community based organisations (NGOs and CBOs) even in slum areas, have come together, often by involving local waste pickers to form durable ventures by which solid waste was collected from households; segregated into bio-degradable and non-bio-degradable waste, its organic element composted; its recyclable element retrieved and the non-compostable non-recyclable residue, a fraction of the original waste, carried to the municipality's collection centre.

The Indian NGO, Exnora International, (an acronym derived from 'excellent', 'novel' and 'radical'), focuses not only on primary collection of waste, street cleaning, vermi-composting of kitchen waste, transportation to the municipal *dhalao* but also tree planting and maintenance of street gardens. The residents of an area (typically a street) form a community based civic unit, Civic Exnora, which organises its own collection and handling of waste generated within the area and cleans and beautifies the area through regular street sweeping and tree planting programme. Exnora, which started in 1989 in Madras with only nine members, had almost 300,000 members by 1996 with over 1,000 civic units in Madras city alone. The garbage is collected from door-to-door by a 'street beautifier', usually an ex-waste picker, employed by Civic Exnora. The collector of waste is called 'street beautifier' not merely to give a positive image to the position but because his activities include both cleaning and beautifying the area. Householders pay for his services and a unit of Civic Exnora is self-sustaining.[37]

37. Venkateswaran, Sandhya (1996). *Alternative Solid Waste Management Systems*, p. 42, National Institute of Urban Affairs, Delhi.

In some other CBO/NGO initiatives, similar ventures were made self-sustaining by evolving a consensus on differential rates of payment for the same service. Innovative methods against those who delayed payment such as a threats of dumping a trolley load of waste at the doorstep were used effectively.[38]

In Latin America, because of the efforts of CBOs/NGOs, waste pickers became trained and confident enough to launch their own cooperatives and micro enterprises which offered waste management services such as street sweeping, recycling, park cleaning and primary collection.[39]

In some of the CBO/NGO initiatives for managing waste, low cost technologies from other low income countries were adapted to suit local conditions. This was the case of the Indonesian Windrow technique of aerobic composting which was adopted by the Dhaka based NGO, Waste Concerns, for its community based decentralised composting pilot project. This programme was also integrated with a waste primary collection initiative. Carbon-nitrogen ratio of 35 to 50 is considered as optimum for aerobic composting. At a higher carbon nitrogen ratio, nitrogen may be a limiting nutrient. To suit the chemical composition of local kitchen waste which had a high carbon content, chicken and cattle manure was added to make up the deficiency of nitrogen. The experience of Waste Concern in decentralised composting in Dhaka proved, after five years of operation, that small-scale composting was commercially profitable.[40] For the purpose of establishing a low-cost and sustainable solid waste management facility for a slum in Dhaka, Waste Concern, used a composting method based on the Sri Lankan model successfully implemented in Colombo. Groups of six households were formed and group members selected 20 locations for placing of barrels for composting of organic waste and four normal barrels for disposal of inorganic waste. Since in low income settlements, major portion of the waste generated is bio-degradable, this distribution was

---

38. *Ibid*, p. 37.

39. Hardoy, Jorge E, *et al., op. cit.*, p. 27.

40. Enayetullah, Iftekhar, A.H. Sinha and Md. Maqsood, "Community Based Decentralised Composting: Experience of Waste Concern in Dhaka" in Enayetullah, Iftekhar, A.H. Sinha and Md. Maqsood, (eds.), *Community Based Solid Waste Management: The Asian Experience*, p. 69-73.

considered adequate. Specially designed perforated green barrels with lids made available for collecting kitchen waste. The slum-dwellers were motivated to dispose their kitchen waste in the green barrels. After three months the aerobically made compost was ready for sale. The proceeds from the sale of the compost were distributed to the participating households to ensure that the experiment endured. The inorganic waste collected in the regular barrels was carried to the nearest municipal dustbin.[41]

Participation of the Private Sector

Advances in bio-technology through which bio-degradation could be expedited with the help of micro-organisms has made composting of segregated organic waste a viable commercial proposition for private entrepreneurs, especially since chemical fertilisers in low income countries have proven ineffective in maintaining high productivity levels in rural areas. The NPK elements (Nitrogen, Phosphorus, Potassium) in the soil are fast depleting due to the large use of animal dung and agricultural residues as fuel. According to a study carried out in Bangladesh in 1998, content of organic matter in soil is estimated at less than 1 per cent whereas the critical level is 3 per cent. 98.69 per cent of the households in Bangladesh use cow dung, husk and wood as fuel. The city made organic manure from segregated waste is now reorganised as an ideal plant nutrient. Many compost-manufacturing plants have been established by the private sector in India—in Bangalore, among other cities.

Small private contractors in cities in North Africa, like Lagos and Ibaden, have begun to collect and transport waste through occuring a contract or selection secured through open bidding conducted by the local government from specified residential areas to designated dump sites suggesting yet another type of initiative by which waste services at the local level could be improved.[42]

In India the activities of the NGO, Sulabh Sauchalaya Sansthan (now known as the Sulabh International Social Service Organisation),

---

41. *Ibid*, pp. 80-81.
42. Ogu, Vincent Ifeanyi (2000). "Private Sector Participation and Municipal Waste Management in Benin City, Nigeria," in *Environment & Urbanisation*, Vol. 12, No. 2 October 2000, p. 85.

have been in a class by themselves. Sensing keenly the inadequacy of the technologically advanced centralised sewerage system in a continental sized country for providing the continuous need for safe excreta disposal, consistent with the cultural requirement of water as a cleansing agent, its founder pioneered the two-pit pour-flush toilet which requires 1.5 litre to two litre of water as against the normal requirement of ten litre in a conventional toilet. Futher, it could be easily installed in the house. It required no municipal worker to handle waste because faecal matter was transformed into a soil like substance on-site which the householder could take out himself and use fertiliser. This toilet could be adapted for use for up to 100 people. The health benefits were further enhanced by the fact that the NGO initiated schooling and vocational training to mainstream children from caste groups associated with cleaning toilets and transporting excreta.[43]

The attitude of municipalities to these ventures was ambivalent. In some limited areas such as primary collection of waste, successful partnerships were established. In Hyderabad, in non-slum areas, Residents Welfare Associations (RWAs) supervised the collection of garbage by the municipal corporation staff on a daily basis because the Corporation subsidised the salary paid to the RWA employee, who regularly collected the waste from each household and placed it in the common collection centre. In addition, the municipality provided the tricycle for collection of waste. However, in extending support to the NGOs/CBOs through regular secondary collection or its regular removal to the landfill site or in the matter of making changes in the design and location of the collection centre or organising of market support for organic compost, the municipality was less accommodating. Its defence was lack of funds, shortage of adequately trained personnel and internal organisational constraints. Besides, in the view of the municipality, people's apathy towards cleanliness of public spaces, was a hindrance. The small scale of a successful NGO/CBO initiative was also regarded as a deterrent to greater cooperation with the NGOs/CBOs. From the municipality's perspective, it meant that such initiatives could not be scaled-up to service vast areas in the municipal jurisdiction where there was greater diversity in social and

43. Sulabh International Social Service Organisation, *www.sulabhinternational.org*

professional background and lower financial and social capability, and lack of certainly of success achieved else where.

The NGOs were not convinced by arguments in the defence of the municipality's status quoist position. Anselm Rosario of Waste Wise, a Bangalore based NGO which has been engaged with SWM issues since the early 1980s noted that a feature of community based centralised waste management systems was that people were willing to pay for services. According to Rosario, "Such an attitude is remarkable in a no pay culture where public sector is traditionally forced to cover the costs. Unfortunately, this trend is not fully activated by authorities."[44] Bindeshwari Pathak voiced similar sentiments because though after 1974 after he demonstrated the viability of a pay and use community toilet in Patna, some state governments/local bodies opted for this system in areas where there was no sewerage system, they were only a handful. By and large there was unwillingness to let go of the familiar septic tank/settled sewerage system installed until such time as the trunk sewer connection was established even though, as Pathak said, in comparison with the two-pit pour-flush system it posed much greater health hazards to lower municipal staff or their substitutes who were periodically required to clean the septic tanks as also to the residents in its vicinity who were unavoidably exposed to pathogens. Other potential savings in his system were costs in transporting and disposing of the sludge. "We often cry ourselves hoarse about the general lack of political will to ensure sanitation for all. Even before political commitment, the mindset of our engineers must change. Only then will this lead to administrative support which in turn will generate the political will and vision in favour of pour-flush technology."[45]

Initiatives to reduce hazardous waste have also not been able to compel government support. Central Pulp and Paper Research Institute in Saharanpur, India had to abandon its pilot plant set up with UN funding to develop 'ferrite technology' meant to assist agro-

44. Rosario, Anselm, "Community Based Decentralised Solid Waste Management Systems: Forging New Partnerships," in Enayetullah, Iftekhar, A.H. Sinha and Md. Maqsood, (eds.), *op. cit.*, p. 51.

45. The Sulabh Movement, p. 15.

based pulp and paper mills reduce toxic effluents generated during production, once the UN funds stopped coming.[46]

Support of National Governments

Both national governments and aid giving organisations have realised that growth in the number of urban poor stretched the municipality's capabilities to provide adequate waste management services. But their interventions have proved too diffused and indirect to ensure durable improvements in the manner in which waste was collected and reduced. An example is the Urban Basic Services Programme of the Indian government which was launched as a pilot project in 1985. The Central government, the State governments and the UNICEF funded the programme. Its object was to help women and children by providing water and sanitation services, immunisation and other community welfare measures. Though subsequently the UNICEF withdrew its financial support from the programme, the Central government, recognising the potential of the scheme, decided in 1990 to implement it nationwide. The programme was renamed as the Basic Service for the Poor (UBSP) and implemented in 245 cities by the mid-1990s. While the UBSP delivered material benefits to many slum communities and helped women realise that they could overcome some of the problems affecting them, there was no improvement in waste related practices due to the absence of linkages with concerned government departments and NGOs. In 1997 the UBSP was discontinued along with two other urban poverty alleviation schemes and replaced by a new scheme the Swarna Jayanthi Sahari Rozgar Yojna.[47] This meant in effect that the supply of whatever meagre funds were available for SSWM ended altoghther.

Role of International Agencies

Table 2.3 shows the proportion of total funding going to urban development and urban infrastructure, urban services and urban management projects of select international agencies. This analysis is

---

46. "Black liquor," (2003). *Down to Earth*, June 15, New Delhi, p. 33.

47. Chaplin, Susan E, *op. cit.*, p. 153.

based on all loan and credit commitments made by these agencies to urban projects.[48]

Table 2.3

Proportion of Total Funding Going to Urban Development and Urban Infrastructure, Urban Services and Urban Management

Funding Org.	*1981-1983*	*1984-1986*	*1987-1989*	*1990-1992*	*1993-1995*	*1996-1998*	*All Years*
World Bank	14.8	17.5	18.7	22.5	27.7	22.1	20.7
Asian Development Bank	21.4	20.3	20.8	22.5	22.6	25.5	22.7
OECF	—	—	20.3	24.3	34.5	39.0	30.1

Note: OECF–Overseas Economic Cooperation Fund of Japan.

The figures above suggest that there has been a steady increase in the proportion of total funding going to urban development and urban infrastructure projects from the early 1980s till the late 1990s. In the case of the World Bank from 1981-1983 to 1993-1995, there was a sharp increase and some decrease in 1996-1998. In the case of the Asian Development Bank, the proportion of total funding rose more or less steadily from 21.4 per cent in 1981-1983 to 25.5 per cent in 1996-1998. In the case of OECF, there was a steady and considerable increase in proportion of total funding between 20.3 per cent to 39 per cent from 1987-1989 to 1996-98.

This is only an indication of the total funds made available to the low income countries in a newly identified area of importance. However, out of the funds made available to the national government, whether and how much a local government received depended upon political considerations. Media attention to a local governmental problem such as a health crisis could clinch the case for a grant. It was also possible there was only a small allocation for SWM in the original allocation meant for upgrading urban services which meant that the number of beneficiaries as also the quantum of money made available could only be low.

Arif Hussain, who pioneered waste management projects for the urban poor in Pakistan, noted in 1999 that over the previous 25 years,

48. Satterthwaite, David (2001). "Reducing Urban Poverty: Constraints on the Effectiveness of Aid Agencies and Development Banks and Some Suggestions for Change," *Environment & Urbanisation*, (April) Vol. 13, No. 1, p. 154.

nine internationally funded studies had been prepared for Karachi to suggest improvements in solid waste management but none of the studies recognised the crucial role of the recycling industry that apart from many economic benefits greatly reduce the amount of waste that needed to be disposed of.[49]

Assessment of the Solid Waste Management Perspective

The limited impact of attempts of civil society groups, committed individuals and international agencies in the 1980s and 1990s to influence local, regional and national governments for the purpose of modifying legal and structural frameworks on more desirable lines has some significant implications for the better management of waste for the future. It is now widely accepted that the urban waste which is growing steadily is not merely food scrap, paper and plastic. It is increasingly heterogeneous and contains not only huge amounts of horticulture waste, fused packaging material, pathogen and infection laded material but also complex chemical compositions. Public health has to be protected not only from bacterial disease such as jaundice, diarrhoea, cholera but also neurological damage, dermatitis, damage to the reproductive system and depression of the immune system. Local environment quality has to be preserved. Even if its recognisable features have to succumb to the pressures of urbanisation and 'concretisation' the pollution levels should be within what the surrounding air, land and water can absorb. The task is too huge for government agencies acting on behalf of a sovereign state to handle by themselves. But their role is crucial in bringing together all social groups in a collaborative relationship to control and regulate waste in a manner which meets the common aspiration for material prosperity and higher consumption level.

Against this background there is a compelling need to reassess conventional wisdom about solid waste management. We need a more inclusive and holistic paradigm. This will necessarily have to be fully sensitive to modern technological developments, sub-national, national and international political power configurations, and participatory modes of decision-making which can encompass all interested groups and governments authorities. All these

49. *Ibid*, p. 139.

considerations were there in the perspective which was spelt out in 1992 in Rio de Janeiro.

The Sustainable Solid Waste Management Paradigm

An alternative paradigm for solid waste management (SWM) for low income countries was proposed in seminal form at the UN Conference on Environment and Development at Rio de Janeiro in June 1992 20 years after the first global conference on the human environment. More than 172 participating countries, the majority from middle and low income countries endorsed it. Hundreds of NGOs from all parts of the world who were present at the conference expressed their support for this new perspective on waste.

A blue print for sustainable solid waste management (SSWM) was part of an exhaustively defined action plan called Agenda 21 meant to operationalise the consensus on conducting the development process in a manner that was more universally acceptable. It was generally felt that the gains of development should be more widely available and that too in a more durable manner. A resolve for such a model of development—'sustainable development' was embodied in the Rio declaration. In shaping the consensus on sustainable development, the 1987 report of the World Commission on Environment and Development (WCED) called "Our Common Future" had a major role. This Commission had been set up by the United Nations General Assembly under the Chairmanship of the Norwegian Labour Party leader, who later became Prime Minister of Norway, Gro Harlem Brundtland. The object of the Commission was to forge a perspective that could guide industrial growth in the coming century. As the authors of the Report put it, "We came to see that a new development path was required, one that sustained human progress not just in a few places for a few years, but in the entire planet into the distant future."[50] The Commission consulted with politicians and bureaucrats from all over the world to arrive at its conclusions. The Brundtland Report defined sustainable development as, "development that meets the needs of the present without compromising the ability of future

---

50. Brundtland, G. (ed.), (1987), *Our Common Future: World Commission on Environment and Development*, p. 4, Oxford University Press, Oxford.

generations to meet their own needs."[51] "Human beings" as the first principle of the Rio Declaration was to put it later were "at the centre of concerns for sustainable development" and they were implicitly, regardless of where, and when they were born "entitled to a healthy and productive life in harmony with nature."[52]

Poverty was an obstacle in the pursuit of sustainable development. Defining poverty as the absence of means to fulfil basic needs, the Brundtland Report stated that sustainable development meant recognition of "the basic needs of all and extending to all the opportunity to fulfil their aspirations for a better life." Poverty drove societies into self-destructive environmental behaviour whose impact was not and nor it possibly could be restricted to them alone. Consequences of use of land, water and air resources by their nature could not be entirely controlled by man-made boundaries. "A world in which poverty is endemic will always be prone to ecological and other catastrophes."[53]

Figure 2.1

Sustainable Development

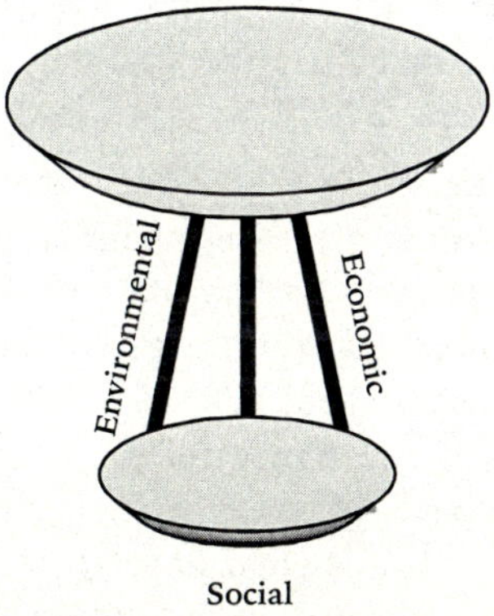

---

51. *Ibid*, p. 8.

52. *The Rio Declaration on Environment and Development* (1992). United Nations Department of Economic and Social Affairs, Division for Sustainable Development, *www.un.org/esa/sustdev*.

53. Brundtland, G., *op. cit.*, p. 8.

The Report went on to say, "We have in the past been concerned about the impacts of economic growth upon the environment. We are now forced to concern ourselves with the impacts of ecological stress... upon our economic prospects. We have in the more recent past forced to face up to a sharp increase in economic inter-dependence among nations. We are now forced to accustom ourselves to an accelerating ecological inter-dependence among nations. Ecology and economy are becoming even more interwoven—locally, regionally, nationally, and globally—into a seamless net of causes and effects."[54]

Sustainable development was 'development' that was required to be economically affordable, socially acceptable and environmentally effective as the Figure 2.1 illustrates. "Equal consideration of each is necessary otherwise the whole system will become unbalanced."[55]

The view of the Brundtland Commission that the developing countries should be assisted to deal with the problem and consequences of poverty was embodied in the fifth and sixth principles of the Rio declaration. Principle 5 stated that, "All States and all people shall cooperate in the essential task of eradicating poverty as an indispensable requirement of sustainable development." Principle 6 indicated that, "The special situation and the needs of developing countries particularly the least developed and those most environmentally vulnerable shall be given priority."

In return, the logic of the Brundtland Report suggested, that the low-income countries needed to remember that as nations who, as the second principle of the Rio declaration stated, had the "sovereign right to exploit their own resources pursuant to their own environmental and development policies," they needed to make wise choices. Their policies and structures to mitigate poverty however radical the thrust, should try to confine any adverse consequences to their own frontiers. Environmental Impact Assessment as a national instrument should be undertaken for activities where a significant adverse impact on the environment was likely (Principle 17).[56]

---

54. *Ibid*, p. 5.
55. McDougall, Forbes, *et al.*, (2001),
56. *The Rio Declaration on Environment and Development*, (1992), *op. cit.*

Three further guiding principles reinforced the consensus on more responsible conduct by states. The first was 'the precautionary principle' that stated, "Where there are threats or serious or irreversible damage, lack of full scientific certainty shall not be used as a reason for postponing cost-effective measures to prevent environmental degradation" (Principle 15) The second was the 'polluter should pay principle,' which stated that "national authorities should endeavour to promote the internalisation of environment costs and the use of environment instruments, taking into account the approach the polluter should, in principle, bear the cost of pollution." Finally, the 'cooperative principle' as embodied in Principles 9 and 10 indicated that "environmental issues are best handled with the participation of all concerned citizens, at the relevant level" and that for endogenous capacity-building there should be improvement of "scientific understanding through exchanges of scientific and technological knowledge," including new and innovative technologies between states.[57]

From the Brundtland Commission Report every state was made aware that pursuit of growth *per se* was not an issue any more. Growth was indeed needed to alleviate poverty and provide basic needs. "Growth has no set limits in terms of population or resource use beyond which lies ecological disaster," but it must be eco-friendly type of growth that is "less material and energy intensive and more equitable in its impact."[58] The same Report also emphasised that the distributive benefits of growth depended upon the number of people who needed to benefit and prevailing consumption patterns. As Principle 8 of the Rio Declaration put it, to achieve a "higher quality of life for all people, states should reduce and eliminate unsustainable patterns of production and consumption and promote appropriate demographic patterns."[59]

For a low income country what did sustainability mean in the context of managing its solid waste, the unavoidable consequence of economic growth/development? Protection of public health and maintenance of safe surroundings in a manner that community could

---

57. *Ibid.*

58. Brundtland, G., *op. cit.*, pages 45, 52.

59. *The Rio Declaration on Environment and Development*, (1992), *op. cit.*

afford has been the traditional goal of SWM. The emphasis of the Brundtland Report and the Rio principles on the parallel pursuits of social equity, environment protection and economic development suggest that for an understanding of SSWM, the three dimensions of SWM, the social, environmental and economic required review and redefinition.

Since health is a basic need of all human beings, apart from regular cleaning of public spaces and keeping them infection free, continuous demonstrated support for livelihood concerns of poorer citizens was required since health and immunity levels are dependent upon nutrition and income. At the level of environment protection, the challenge was to match 'less material intensive production' with fewer emissions on land, air and water. The 'carrying capacity' of local water bodies to absorb untreated sewage safely would have to be a relevant consideration for public authorities. Further, since natural resources were limited and economic and environmental costs of production from recycled material were lower than that from virgin material, recovery of natural material from waste was also an objective required by the environmental dimension. In the SSWM perspective the operational costs of the services that were provided needed to be considered as being much higher than in SWM. Though the cost of undoing the cumulative impact of pollution could be staggered over some years in the future, 'polluter must pay principle' suggested that whatever pollution was caused in the present by a territorial unit would have to be paid for by its existing members. A differential rate reflecting payment capacities could be adopted given the unequal income levels. In this manner the concept of SSWM can be said to embody the two principles of intra-generational equity and inter-generational equity that are a characteristic of the Brundtland Commission's definition of sustainable development and possibly the most widely quoted definition in situations where the term has relevance.

Significant features of the SSWM perspective can be identified from Chapter 20 and 21 of Agenda 21 that deal exclusively with waste related issues. Chapter 20 is entitled the 'Environmentally sound management of hazardous waste, including prevention of illegal international traffic in hazardous wastes' and Chapter 21 is called 'Environmentally sound management of solid wastes and sewage related issues.' Due to their common potential of adversely impacting on the environment and society wastes from different sources are discussed together in Agenda 21.[60]

Sources and Characteristics of Waste in SSWM

Sources of solid waste as defined in Chapter 21, include "all domestic refuse and non-hazardous wastes such as commercial and institutional wastes, street sweepings and construction debris," and waste from "domestic and transnational industry." In some countries, it is noted the solid wastes management system also handles "human wastes such as night-soil, ashes from incinerators, septic tank sludge and sludge from sewage treatment plants." (Para 20.1 and Para 21.3). Wastes are defined as non-hazardous and hazardous and while household, commercial and non-commercial institutions, and street sweepings are considered as sources of largely non-hazardous waste, night soil component in waste collected by the SWM system is to be regarded as hazardous if it manifests these characteristics. The explicit recognition of faecal matter as hazardous and its discussion under a chapter largely concerned with non-hazardous waste even as toxic chemical waste is elaborated upon in the very next chapter suggests that in the SSWM perspective waste is holistically defined and that all wastes from households institutions and factories enjoy parity. Though human excrement is acknowledged as being a component of street sweepings because of its relatively miniscule proportions of the total waste its implications for the management system are ignored in the conventional SWM perspective. 'Night-soil,' 'liquid waste' or 'black water' and related social and environment impacts are considered adequately addressed by a different governmental agency's even when mainstream non-hazardous waste management suffers due to the presence of excreta among other factors. A similar *laissez faire* approach is adopted in SWM to deposition of industrial hazardous waste in unauthorised open spaces within municipal limits. A distinct feature of SSWM is that it requires non-hazardous waste, faecal matter and toxic chemical waste from domestic and multi-national industry to be treated together as aspects of a single solid waste problem for whose resolution concerned governmental agencies are required to coordinate their functioning.

---

60. *Agenda* 21, United Nations Division for Sustainable Development, *www.un.org/esa/sustdev/agenda 21.*

Waste Managing Authorities in SSWM

Unlike the conventional solid waste management perspective that focuses on municipalities and local governments, in Agenda 21, SSWM is addressed to national governments that are advised to recognise the significance of the smallest administrative unit for the purposes of waste management. "Local authorities construct, operate and maintain economic, social and environmental infrastructure, oversee planning processes, establish local environmental policies and regulations, and assist in implementing national and sub-national environmental policies. As the level of governance closest to the people, they play a vital role in educating, mobilising and responding to the public to promote sustainable development." (Para 28.1).

The importance of the central government is emphasised for its competence in constructing comprehensive and widely applicable legal frameworks and organising and facilitating technical assistance while providing incentives to industry to assume responsibilities for the waste it generates. Agenda 21 stated that policy changes at the national and local levels should include "giving full recognition to and using the full range of low-cost options for waste management, including, where appropriate, their institutionalisation and incorporation within codes of practice and regulation," and "assigning high priority to the extension of waste management services, as necessary and appropriate, to all settlements irrespective of their legal status, while giving due emphasis to meeting the waste disposal needs of the unserved, especially the unserved urban poor." (Para 21.45).

National governments should seek support of other governments and inter-governmental institutions to manage their waste in a sustainable manner. In connection with the management of hazardous waste, Agenda 21 state, "governments, according to their capacities and with the help of multilateral cooperation, should provide economic or regulatory incentives, where appropriate, to stimulate industrial innovation towards cleaner production methods, to encourage industry to invest in preventive and/or recycling technologies so as to ensure environmentally sound management of all hazardous wastes, including recyclable wastes, and to encourage waste minimisation investments." (Para 20.13b)

Waste Management Operations in SSWM

In the conventional SWM approach for ensuring durable freedom from non-hazardous waste, the focus is on the distinct tasks of collection of the estimated quantity of waste, its transportation, possible treatment, composting and disposal. Though the waste management hierarchy principles are widely acknowledged and praised, they are considered impractical and peripheral to the basic mainstream activity of regular removal and movement to safer destination of thousands of tonnes of the waste generated within city limits every day. In SSWM these principles are required to be incorporated in the functioning of government agencies concerned with the management of all the three main categories of solid waste—non-hazardous, human and hazardous.

In the context of the first type of waste, Agenda 21 government departments would have to, "go beyond the mere safe disposal or recovery of wastes that are generated and seek to address the root cause of the problem by attempting to change unsustainable patterns of production and consumption," and "application of the integrated life cycle management concept" or cradle-to-grave approach towards the waste generated.

About non-hazardous waste, the prediction in Agenda 21 was that in 2025 at existing rates of growth it would be four to five times than what it was in 1992. "A preventive waste management approach focused on changes in lifestyles and in production and consumption patterns offers the best chance for reversing current trends" (Para 21.7). For this purpose, "regulatory and non-regulatory incentives to encourage industry to change product design and reduce industrial process wastes through cleaner production technologies and good housekeeping practices and to encourage industries and consumers to use types of packaging that can be safely reused" were advised (Para 21.14).

Since non-hazardous waste originated from homes and offices, it had substantial amount of organic food waste and other paper, plastic, metal and glass. The waste hierarchy's imperatives were that these should be source segregated and subjected to composting, recycling and waste recovery. Agenda 21 recommended increasing funding for research pilot programmes to test various options for reuse and recycling, including the use of small-scale, cottage-based recycling industries; compost production; treated waste-water irrigation; and energy recovery from wastes (Para 21.20 c). Further, potential markets for recycled products needed to be identified. (Para 21.20 f) Agenda 21 explicitly recognised the synergy that could be generated by bringing together the waste-pickers and the affluent waste generating householders. There could be provision for "technical assistance to informal waste reuse and recycling operations" and application "of economic and regulatory instruments, including tax incentives, to support the principle that generators of wastes pay for their disposal" (Para 21.24 b and c).

If existing municipalities were to apply the principle of SSWM as envisaged in Agenda 21 they would have to modify prevailing expenditure on vehicles and personnel deployment patterns so as to accommodate new infrastructure such as waste sorting and composting sites close to the point of waste generation, and organise door-to-door collection of waste. They would be required to establish durable partnerships with community based organisations, non-governmental organisations and the private sector and forge links with specialised institutions so that staff members enhance skills in finance information system to 'build capacity' as it were, and to understand the importance of 'attitudinal changes.' Appropriate policies and structural changes involving institutions, finance, and technology were unavoidable requirements of SSWM. Agenda 21 stated that governments, in collaboration with United Nations international and regional organisations, should undertake actions including such as "waste reuse and recycling in in-service training programmes as integral components of technical cooperation programmes on urban management and infrastructure development" and encourag non-governmental organisations, community-based organisations and women's, youth and public interest group programmes, in collaboration with local municipal authorities, to mobilise community support for waste re-use and recycling through focused community-level campaigns. (Para 21.25)

About hazardous human waste, Agenda 21 expressed the view, that due to inadequate sanitation facilities in the developing countries it posed a serious hazard to public health and besides being a pollutant of freshwater resources. Less than 10 per cent of urban wastes received some form of treatment and only a small proportion of treatment was in compliance with any acceptable quality standard. "Faecal matter treatment and disposal should be accorded due priority given the potential threat of faeces to human health... As many as 5.2 million people, including four million children under five years of age, die each year from waste-related diseases. The health impacts are particularly severe for the urban poor" (Para 21.27 and Para 21.38).

The application of waste hierarchy to human waste meant all such waste must be discharged into land or water so that it does not pose a threat to life forms. The technology was irrelevant. However, if the technology imposed a burden on the exchequer than its beneficiaries should pay. Agenda 21 stated that the "health and environmental impacts of inadequate waste management go beyond the unserved settlements themselves and result in water, land and air contamination and pollution over a wider area. Extending and improving waste collection and safe disposal services are crucial to gaining control over this form of pollution," and suggested the application of the "polluter pays" principle, where appropriate, by setting waste management charges at rates that reflect the costs of providing the service and ensure that those who generate the wastes pay the full cost of disposal in an environmentally safe way" (Para 21.38 and Para 21.40).

Though the level of chemical waste generated by industry was far less than that of human waste, its management until the moment of final disposal and the method of disposal itself posed a huge challenge to low-income countries because of lack of adequate information about its adverse impacts on health and environment and waste disposal sites which were expensive to construct. According to Agenda 21, "effective control of the generation, storage, treatment, recycling and reuse, transport, recovery and disposal of hazardous wastes is of paramount importance for proper health, environmental protection and natural resource management, and sustainable development. This will require the active cooperation and participation of the international community, governments and industry" (Para 20.1).

The problem of hazardous waste in low income countries was accentuated due to the fact that in defiance of national legislation and international conventions, hazardous waste was being dumped in developing countries by the developed countries. The first step in hazardous waste management was ratification of the Basel Convention on the Control of Transboundary Movements of Hazardous Wastes that made it possible to sue or take to court a developed country for illegal dumping of hazardous waste. Environmentally sound hazardous waste management required that as far as possible it should be disposed of within the country of its origin (the self-sufficiency principle). The "transboundary movements that take place should be on environmental and economic grounds and based upon agreements between the States concerned" (Para 20.7).

According to Agenda 21 the application of the waste management hierarchy to hazardous waste suggested three goals: "To reduce the generation of hazardous wastes, to the extent feasible, as part of an integrated cleaner production approach; to optimise the use of materials by utilising, where practicable and environmentally sound, the residues from production processes; and to enhance knowledge and information on the economics of prevention and management of hazardous wastes" (Para 20.11).

"Further, the endeavour of governments should be to intensify research and development activities on cost-effective alternatives for processes and substances that currently result in the generation of hazardous wastes that pose particular problems for environmentally sound disposal or treatment. The possibility of ultimate phase-out of those substances that present an unreasonable or otherwise unmanageable risk and are toxic, persistent and bio-accumulative should be considered as soon as practicable and there should be cooperation with industry to develop guidelines and codes of conduct, where appropriate, leading to cleaner production through sectoral trade industry associations" (Para 20.13).

Agenda 21 also called upon the governments to:

- "Encourage industry to treat, recycle, reuse and dispose of wastes at the source of generation, or as close as possible whenever hazardous waste generation is unavoidable and when it is both economically and environmentally efficient for industry to do so" (Para 20.13).
- "Establish regulations that lay down the ultimate responsibility of industries for environmentally sound disposal of the hazardous wastes their activities generate" (Para 20.13).

- "Encourage industries to be transparent in their operations and provide relevant information to the communities that might be affected by the generation, management and disposal of hazardous wastes" (Para 20.14).

Improvements in the Waste Management System

Perhaps the most striking difference between the SWM and SSWM perspectives concerns the end result of governmental initiative directed to this goal which necessarily is the same, *viz.* making an area garbage free on a regular basis. In SWM, the results can be ensured since availability of adequate capital and discretion to senior government decision-makers are the critical inputs required for the purpose of making waste invisible by shifting it to another site. In SSWM 'cleanliness' is regarded as a by-product of systemic change, a complicated process which requires tedious preparatory work besides time consuming efforts for creating appropriate broad-based social, physical and legal-political infrastructure. Klundert and Anschutz who have devised a three dimensional Integrated Solid Waste Management (ISWM) model for low and middle income countries say that in connection with its application for municipal non-hazardous waste, the preliminary step, to merely 'know' the waste, requires four steps.

- **Know the source and composition of waste**. It is important to know who is generating what type of source, in what quantities and where. The composition of waste partly determines the suitability of the waste for certain type of treatment.
- **Know the city and relevant neighbourhoods**. It is necessary to know about the physical infrastructure of the city such as road and traffic conditions, types of buildings and spaces for storage of waste because these conditions influence choice of collection and transportation technology and may also influence the collection routes and opportunities for recycling.
- **Know the citizens and clients of the waste system**. Their ability and willingness to cooperate in the operation and management of the service will be crucial to the success of the system.

- **Know all waste management activities in the city.** It is extremely important to know about such activities which include identifying who is doing what in the formal and informal waste management systems and involving stakeholders in assessing the performance of the waste management by the municipality, community groups and large, small and micro-scale enterprises. The performance assessment could take the form of SWOT analysis (exploring strengths, weaknesses, opportunities and threats of the various players).[61]

In the first dimension of the model, the focus is on forging links with stakeholders-persons or organisations that have an interest in SWM. A typical list will include, they say, apart from the municipal workforce, the informal sector waste-picker, itinerant waste buyer, the dealer who buys, sorts and sells materials, the wholesaler who usually specialises in one type of material and sells it to the industry, the recycling enterprise that processes recyclables into intermediate industrial feed stocks and the end-user industries which purchase processed recyclables to make their final product. In the second dimension, the waste system elements are fore grounded and the traditional collection, transfer and disposal functions are by and large subordinated to prevention/minimisation, reuse and recycling. The final element consists of six sustainability aspects or lenses reflecting local conditions by which the existing waste system is assessed and a new system planned. These are environmental aspects which focus on pollution control and resource recovery; political/legal aspects that deal with the regulatory decision-making framework; the institutional aspects which relate to the political and social structures that control and implement waste management; the socio-cultural aspects which include the influence of culture on waste generation and management in the household and businesses and institutions; financial and economic aspects that pertain to budgeting and accounting within the waste management system and relations to the local, regional, national and international economy. Where some specific issues such as privatisation, cost recovery and cost reduction and connection of the recycling infrastructure to the market place can be addressed.[62]

---

61. Van de Klundert, Arnold and Anschutz, Justine, *op. cit.*, p. 18.

62. *Ibid*, pp. 12-14.

According to Klundert and Anschutz, development and planning are long-term issues, which require time to occur and to mature. Foreign donor agencies and local decision-makers do not always realise this, which leads them to take *ad hoc* decisions or propose short-term projects that reflect well on their term in office. Local or national elections may also lead to reversal of policies and obstruction of attempts to arrive at long term but sustainable solutions. It also takes time to learn new habits and unlearn old bad ones. It may take time before new attitudes and behaviour regarding, for example, separation at source is rooted in society.[63]

Klundert and Anschutz's views which suggest that there is no single simple indicator of the success of institutionalisation of SSWM are in line with the insight of the Brundtland Report. The Report says, "sustainable development is not a fixed state of harmony, but rather a process of change in which the exploitation of resources, the direction of investments, the orientation of technological development, and institutional changes are consistent with future as well as present needs."[64]

Assessment of the Sustainable Solid Waste Management Paradigm

Thus in 1992, the governments of low income countries had become aware a new waste management paradigm, which was sensitive to their economic conditions and offered the promise of cleaner and healthier surroundings. In the Indian context this was particularly significant. Where Article 48A of the Indian Constitution, explicitly stated that the State shall endeavour to "protect and improve the environment." The new perspective demanded a radical shift in thinking but the resources and skills it required were adequate in most states to make a beginning. The pertinent question is why has there been such limited evidence of 'political will' towards its adoption and implementation. An explanation can be found in the political antecedents of the Rio conference; ambiguities and contradictions of the SSWM principles; and constraints rooted in the historical background of the low income countries.

---

63. *Ibid*, p. 14.

64. Brundtland, G., *op. cit.*, pp. 8 and 9.

Political Antecedents of the Rio Conference

SSWM was a dimension of a new concept of development proposed by high income countries to low income countries to bridge the gap between themselves and the developing countries on the contentious issue of restraint in the pursuit of fossil fuel based production. The industrialised countries were acutely aware that with their advanced technologies, in spite of their limited population, they were consuming almost 20 times as much of the environment resources. Gro Brundtland noted in 1992, "An average person in North America consumes almost 20 times as much as a person in India or China, and 60 to 70 times more than a person in Bangladesh."[65] Since, the 1972 Stockholm Conference, their fears about the growing population in the developing countries and a perception of the finite character of natural resources as a threat to the continuation of their lifestyles had become accentuated by new worries about climate change, ozone depletion and biodiversity loss. To the countries in the South, beset with virtual stagnant economic growth, debt crisis, collapsing commodity prices, and growing poverty, any talk of restraint, despite the stark evidence of environmental degradation, was either an anathema or an irrelevance. The easing of the East-West tensions associated with the Cold War until the late-1980s further led to the foregrounding of environmental concerns. Against this background the concept of "sustainable development' was evolved.

The right of each country to use its resources to meet the basic needs of its citizens was conceded and the developed world expressed its readiness to help in making this possible by making available both money and expertise. In return it was required that the physical constraints of the local ecological systems be recognised. Agenda 21 expressed a perspective of the challenges the developing world faced in pursuing the goal of growth and prosperity which was extrapolated from the experiences of the Western world in recent years rather than any close familiarity with ground realities in low income countries. This is clearly evident in the substantive suggestions of Chapters 20 and 21. In connection with reduction of non-hazardous waste a major emphasis was on recycling, an area in which industries in the west, due to state support had acquired considerable expertise in the wake

65. Brundtland, G., quoted in Dresner, Simon, (2002), *The Principles of Sustainability*, p. 88, Earthscan Publications Ltd., London.

of the oil crisis. There was no recognition of the need for safe final disposal of the waste that cannot be composted or recycled. In low income countries, the general practice of open dumping leads to health hazards from leachate and landfill gases. The industrialised countries guard against these through engineered landfill sites. Further, Agenda 21's insightful observation that human waste was hazardous and required a distinct strategy for safe disposal was not matched by an understanding of the extremely inadequate nature of centralised sewer-based waste disposal system imported from the West.

There was no urgency in the high income countries about implementing Agenda 21. No legal sanctions were provided for enforcing the principles. They were looking for policy orientation over the long term. The low income countries had no hesitation in formally accepting the principles because they came with the promise of a bounty. All countries, except the United States, agreed in principle to the aid target of 125 billion US dollars every year for the implementation of the Agenda 21. However, the industrialised countries were not prepared to put forward that kind of money and eventually came up with two billion US dollars over a period of three years, about 0.5 per cent of the sum asked for.[66] It was therefore business as usual for the leaders of the low-income countries once they returned from Rio.

The governments in the low income countries have undertaken review of existing SWM practices and introduced changes largely when there was a public outcry against a perceived health crisis as was the case in Surat in 1994. There were some dramatic improvements but after a while *status quo ante* was restored because there was no policy or fund allocation for systemic change.

Ambiguities in the SSWM Model

SSWM endorses conflicting policies in some cases and the end-result of its application can be confusion resentment. In Chapter 20 of Agenda 21, in relation to the objectives of waste reduction, resource recovery and energy conservation (less energy is used in manufacture from recycled material rather than from virgin material), recycling activities are considered very deserving of encouragement.

---

66. *Ibid*, p. 42.

However, there is no restriction on the import of non-hazardous waste unlike in the case of transboundary movement of hazardous waste. Import of non-hazardous waste from industrialised countries is preferred for recycling by industries such as paper industry since imported waste can be relied upon for quality, quantity and regular availability. In a market driven economy, there is no incentive for adapting machines to use indigenous alternatives, such as agricultural residues for the paper industry. Consequently, quantity of local agriculture waste for disposal through incineration is not reduced. According to the Indian Agro and Recycled Paper Mills Association, India produces about 15 to 20 million tonnes of rice and wheat straw, which can yield about three to four million tonnes of paper annually (about six million tonnes of paper is manufactured in India annually). Another implication of unregulated import of used paper is that it impacts adversary on the income of waste-pickers and the work load of the municipality. There is less incentive to gather paper for recycling because its purchase price goes down whenever large shipments of used paper arrives. Further, the quantity of soiled paper which has to be collected, transported and disposed by the municipality increases. If a decision-maker inspired by considerations of greater social equity and environment protection decides to cut down on imports of recycled paper he can expect to be the object of wrath of both management and workers of an organised sector, who can be expected to mobilise adequate political pressure to have him transferred.

For implementation of SSWM reliance is placed on the government establishing partnerships with other governmental and non-governmental bodies and private sector. Partnership is a value neutral concept. Its results can disappoint if a durable redistribution of power is the intention. As Jewson and MacGregor put it, "Partnerships can also become an attractive concept to government because 'it diffuses responsibility for success or failure' and could become a system for co-opting institutions into an extended system of repressive control." With so many actors and interests involved in a particular partnership, they say, it is highly unlikely that all will come out as winners. "Therefore crucial questions to be addressed in research are: which interests and which players will be included in partnerships and which will be left outside? Who will be the leaders within partnerships? Whose agendas will prevail?"[67] Various entities in a partnership have their own priorities and whether a trade-off will advance the goals of SSWM will depend upon the thinking of the government of the low-income country, which determines the framework of negotiations and also the choice of partners, and whether there are regulations in place to ensure transparency. Bi-lateral and multi-lateral loan and aid giving agencies are known to favour capital intensive technologies and in the absence of adequate safeguards built into the procedures to protect the interests of the recipient body, a generous loan from a foreign bank can be means of gratification for the civil servant besides being a disservice to the cause of SSWM. It could even be a set back for the SSWM if (procurement of waste transport) vehicles is of inappropriate design as many studies have shown.[68]

The inclusion of an indigenous expert in the negotiating team on the government side can represent a huge saving and a cost effective choice for meeting the solid waste management goals. In Pakistan, a 70 million US dollar loan from the Asian Development Bank for part of the Greater Karachi Sewerage Plan was not accepted by the city authorities even though the federal government and the provincial government were in favour of it because the local NGO on board as a consultant could convincingly show that with local resources and expertise a similar project could be prepared at a fraction of the proposed cost. In the process a large debt obligation was avoided. However, it is rare for cities in low-income countries to have local NGOs which have the technical capacity to question the design and proposed budgets for large, complex infrastructure projects.[69]

Agenda 21 while acknowledging the institutional limitations of low-income countries had expressed confidence that through capacity building and human resource development not only could the low

---

67. Jewson and MacGregor (1997), as quoted in Elander, Ingemar, and Lidskog, Rolf, "The Rio Declaration and Subsequent Global Initiatives" in (eds), Nicholas Low *et al.* (ed) *Consuming Cities: The Urban Environment in the Global Economy after the Rio Declaration* p. 35.

68. van de Klundert, Arnold and Justine Anschutz (1999), "Integrated Sustainable Waste Management: The Selection of Appropriate Technologies and the Design of Sustainable Systems is not (only) a Technical Issue," CEDARE/IETC Inter-Regional Workshop on *Technologies for Sustainable Waste Management*, p. 4, Alexandria, Egypt.

69. Satterthwaite, David, *op. cit.*, p. 139.

income countries stop hazardous waste at their frontiers, they could also minimise and safely manage both hazardous and non-hazardous waste in future years. These hopes have been largely belied. In the case of India, as late as 1999, two years after the Supreme Court banned the import of hazardous wastes, these wastes continued to enter India. Though India had acceded to the Basel Convention on Transboundary Movement of Hazardous Waste and their Disposal, the Central government was unable to put in place the infrastructure to enforce the ban. The Central government also failed to take up the issue of illegal exports to India with the exporting countries.[70] In recent years, developing countries have been unable to regulate unauthorised deposit of second hand electronic goods which developed countries find cheaper to 'donate' to destinations abroad rather than dispose of safely as per their own standards. India has the Hazardous Waste (Management and Handling), Rules 1989 but the Rules do not cover e-waste. A contributory reason for this situation is that some of the biggest generators of e-waste like the United States have not ratified the Basel Convention.[71]

History Related Constraints

The pursuit of urban sustainable waste management in the low-income countries faces problems which have no parallel in developed countries because forging durable consensus on national issues can be a source of special difficulty in many low-income countries because of their history as post-colonial societies.

Cleaning related activities in these countries are considered low status and performed by groups which belong to the poorest sections. India, where social hierarchies were based on occupation and perpetuated through heredity, the lowest level associated with 'polluting,' activities like cleaning of private toilets and sweeping of public spaces and skinning of dead animals suffered not only from poverty but also social humiliation and exclusion. Direct rule over India lasting more than a hundred years of the then most politically advanced nation did little to change this reality. After Independence,

70. Jayaraman, Nitynand (2002). "Hazardous Wastes", *The Hindu Survey of the Environment 2002*, p. 51.
71. Jamwal, Nidhi (2003). "e-Waste, Developing countries are Dumpyards for New Millennium Waste", *Down to Earth*, Vol. 12, No. 12, November 15.

the Constitution barred the practice of 'untouchability' and preferential treatment was given to specific caste based ('scheduled') groups in the matter of public employment to compensate for patterns of deprivation in the past. In present day municipalities, substantial number of the workforce at the lowest level is from these groups. Though their productivity is low and benefits and wages high, as compared to similar skill levels in the private sector, the general perception is that their labour unions cannot be expected to agree to any reduction in personnel levels.[72] Even a change in the nature of their duties is resented. A project to implement the Solid Waste (Management and Handling) Rules, from January 2004 in Delhi, was put on hold, until after the December 2003 election, due to employee opposition to the new systems being contemplated for managing municipal solid waste. For all politicians, the municipal union spokespersons are *de facto* leaders of communities whose votes matter. Yet savings resulting from reducing the numbers of regular workforce and the flexibility resulting from their exclusion from certain work areas may be a necessary condition for the implementation of SSWM principles.

In 1999, the Barman Committee after carrying out a nation-wide survey of municipalities was of the view that their dismal condition was due to the perpetuation of priorities inherited from a colonial government. The Report States, "The SWM service has remained extremely neglected in the country ever since-Independence. It never caught the attention of decision-makers. The service has thus remained primitive and inefficient."[73] The Report further observed, "Most of the city fathers and senior decision-making bureaucrats do not consider the subject worth their personal attention. Inspite of this service consuming the largest Municipal budget, the subject is left in the hands of junior level supervisors having no vision, foresight and training."[74]

With the coming into force of the Constitution (74th) Amendment Act, 1993, which provided for elected bodies at the municipal level, a major recommendation of Agenda 21 was in place. In the city of

72. Coad, Adrian, (1995) "Editorial" in *Lessons From India in Solid Waste Management*, Sponsored by the Department of International Development, Government of UK, p. 9.

73. Report of the Committee constituted by the Hon. Supreme Court of India, *op. cit.*, p. 21.

74. *Ibid*, p. 21.

Delhi, this means that since there are 134 Councillors, who get an allocation of Rs. 50 lakh each, Rs. 6,700 lakh is available for expenditure at the ward level as part of the capital expenditure of the MCD which is allotted to the councillors for the development of their wards. With the existing 68 State level legislators and six national level legislators getting Rs. one crore each, a huge sum of Rs. 141 crore is available with the elected leaders for basic services at the local level. With accumulated garbage such an omnipresent problem, one would have imagined that construction of user friendly waste bins, improved waste collection schemes involving NGOs and garbage reducing, profit making compost generating schemes would have received a boost. However, there is no remarkable difference. In 2000 the study carried out on behalf of the Institute of Social Science, Delhi, noted that not many councillors "identified a scheme for SWM, like construction of *dhalaos* in their locality or helping the NGOs with initial capital expenditure for a SWM project."[75]

Clearly, however desirable democratisation in local government may be, it seems irrelevant in the context of dislodging inherited mind sets about waste, a pre-condition for the successful implementation of SSWM. The Constitution (Seventy-fourth Amendment) Act requires the Central government to take appropriate steps to empower local bodies to discharge their new constitutional responsibilities which according to the Twelfth Schedule of the Constitution include urban planning and regulation of land use. In the case of Delhi, the regulation of land use has not been transferred to the city government. The municipality cannot often build public toilets where they are needed most because the land on which the slum stands is owned by some agency of the Central government. Post-Independence authorities have found it difficult to let go of colonial mentality of central control. A further legacy of centralised bureaucratic control in the low income countries is the perceived low importance of the municipal administration. Skilled managers are not usually attracted to the post of Municipal Commissioner. In India, except in the mega cities, it lacks status and is a less attractive position than what a potential incumbent may have been eligible for in other departments of the government.[76]

---

75. Ghosh, Archana (2000). *Solid Waste Management in Delhi*, Institute of Social Science, New Delhi, p. 43.
76. Plummer, Janelle (1999). *Municipalities and Community Participation*, p. 121, Earthscan, London.

Due to the hierarchical nature of municipal institutions, a skilled and committed reform minded official can create a supportive environment to enable the staff to relate with local communities and mobilise their cooperation and the end-result could be dramatic and durable improvement in the SWM conditions. Such was the case in Surat eight months after the outbreak of suspected plague when the IAS officer S.R. Rao took over the post of its Municipal Commissioner. Management reforms introduced were essentially two-fold. The first step was the delegation of powers and the second was the strengthening of the management core into an accountable and responsible team.[77] Despite official commitment to decentralisation and participatory principles in which the SSWM is grounded and examples of impressive delivery by individuals who have held this office, the potential of this position does not seem to have been tapped by posting officers who have the aptitude and skills for this post.

Practical Difficulties

There can be immense difficulties in implementing SSWM principles. It can be that the senior officials of two divisions of the local government agency such as that of horticulture and solid waste may not be equally interested in the execution of a SSWM project which requires their cooperation such as in the case of purchase of compost made from biodegradable waste. Even when many senior and middle level managers support SSWM principles, it may be considered demeaning to show deference to NGOs who have a major role in building municipal capacity by acting as mediators between the municipalities and the community. NGOs can help to facilitate initial discussions, develop community confidence and provide a supportive framework for community action. NGOs can also guide households integrate source segregation into their lifestyles, overcome their distrust of dealing with former waste-pickers as service providers and defuse anxiety about odour and pests due to composting in the vicinity. However, such activities can be peripheral to the NGO's established agenda or the NGO may not be in a position to upscale its work as may be required by the municipality. In poorer communities, entrenched spokesmen who are usually males divided by

---

77. *Ibid*, p. 137-38.

loyalties to various political factions, may be unwilling to cooperate with the municipality in planning for setting up a service delivery programme. This could be because they are sceptical due to the municipality's history of unfulfilled promises or that they are accustomed to being only beneficiaries. Further, at the lowest level, the municipality's own staff, due to a long professional career characterised by a top-down prescriptive procedure, may be unable to listen and relate with patience to ordinary citizens which are necessary qualities for encouraging attitudinal change among those accustomed to being passive recipients of service.[78]

A further difficulty in implementing SSWM which envisages close coordination between governmental and non-governmental bodies is the perception of excessive favourable media attention to 'alternate' community based experiments in waste management. Adrian Coad, Consultant, Swiss Centre for Development Cooperation in Technology and Management (SKAT) who studied SWM operations of Mumbai in the mid-1990s said, "these (pieces) are largely written from a sociological point of view" and "sometimes give the impression that the author is unaware that there is a large municipal organisation which is trying to solve the solid waste problems in the city in question."[79] Senior government officials could feel inhibited about taking up a bold initiative for SSWM due to their uninspiring public image and the low assessment of their capabilities to significantly change bureaucratic systems.

Strengths of SSWM

Views of Scholars

Despite its ideological deficiencies, ambiguities, flawed assumptions and insensitivity towards social constraints imposed by historical backgrounds, scholars of SWM in developing countries have been quick to grasp the potential of the SSWM paradigm in redefining the challenges before SWM in these countries and identifying workable solutions which could result in transforming existing arrangements and ensure broad favourable impacts on health and the environment.

---

78. *Ibid*, pp. 75-77, 121-124.

79. Coad, Adrian, "Editorial" in Adrian Coad (ed) (1995) *Lessons From India in Solid Waste Management*, *op. cit.*, p. 7.

Schubler noted that the essential condition of sustainability implies that waste management system must be absorbed and carried by the society and its local communities. The system must, in other words, be appropriate to the circumstances and problems of the particular city and locality, employing and developing the capacities of all stakeholders, including the households, and communities requiring service, private sector enterprises and workers (both formal and informal), and government agencies at the local, regional and national levels. The success will depend, "as much upon organisation and cooperation between households, communities, private enterprises and municipal authorities as it does upon the selection and application of appropriate technical solutions for waste collection, transfer, recycling and disposal."[80]

On the basis of their study of solid waste management practices in Asia and Africa, van de Klundert and Anschutz, constructed a model of Integrated Sustainable Waste Management which could be applied to solid and liquid waste management in low and middle income countries in the South, and also in countries in transition in East Europe. In their view, the capital and technology intensive and labour extensive models of waste management involving collection with compactor trucks and waste disposal with incinerators, however flawed, have relevance for industrially advanced countries because they are suited to local conditions. The Integrated Solid Waste Management (ISWM) insight is that most waste management problems are not related to lack of resources or equipment but with something other- or more- than money and equipment. "Some problems have to do with the attitude and behaviour of citizens, waste management staff, private enterprises and waste-pickers. Other problems are caused or made more serious by factors that are not technical or financial, but relate to managerial (in)capacities, the institutional framework, the environment, or the social or cultural context. In these cases, it is not money or equipment that provide solutions, but rather changing social, institutional, legal or political conditions."[81]

---

80. Schubeler, P. *et al.* (1996), "Conceptual framework for Municipal Solid Waste Management in Low-income Countries", World Bank Urban Management Programme, *Working Paper No 9, worldbank.org/publications/*, p. 19.

81. Van de Klundert, Arnold and Justine Anschutz, *op. cit.*, p. 10.

Final disposal of waste is the most neglected aspect of municipal solid waste. Even when waste minimisation and recycling are actively practiced, there is always a large quantity of waste remaining for disposal in an environmentally sound manner. The most common method of disposal in developing countries is open dumping which contravenes all accepted landfill principles with regard to siting, design and operation. In this connection, Schubeler has suggested that the "main benefits of properly designed and correctly operated sanitary landfills derive from the discontinuation of current, unacceptable dumping practices and the environmentally sound closure and recovery of existing dump sites. It is seldom possible to move from open dumping to fully contained sanitary landfill operations in one step. More often, a transformation process must be foreseen in which dumping practices are progressively improved and existing sites gradually upgraded."[82] Appropriate consideration of local conditions could provide for a planned upgrade as the necessary finance becomes available.

Many analysts and entrepreneurs (Robert Svadlenka; Otter Wasser gmbh) are in agreement that the 'modern' water intensive (flush) human waste collection and treatment system that requires a network of pipes and pumps connected to a high technology central sewerage treatment plant needs to be phased out. This was not only because it was too costly to meet the needs of a growing population, it was unintendedly contributing to the spread of disease and blocking a cost effective method for increasing agricultural productivity. In China, Indonesia and India, twice as many people died of diarrhea than HIV/ AIDS. It was estimated that parasitic worm infestation which at any time afflicted 1.5 billion people was contracted from human excreta in their surroundings. In using drinking quality water to flush and transport human excreta, not only was a precious and limited resource being wasted, two major soil nutrients, phosphorus and nitrogen, in addition to other organic substances were diluted and discarded into rivers and the sea where they were causing great harm (eutrophication) even as fossil resources were being applied to produce these very nutrients for agricultural purposes.

---

82. Schubeler, P. *et al.*, (1996), p. 48.

The two alternative approaches generally considered appropriate for poor urbanised areas the 'settled sewerage' and the 'simplified sewerage' systems had many limitations from the point of view of meeting mass needs in resource poor societies. In the first, the solid portion is intercepted by a septic tank located adjacent to the home/homes. The limitation of this system is that it required a centralised sanitation authority with command over a technology that could regularly and safely remove the sludge from the septic tanks. This approach also required a liquid waste treatment facility like a stabilisation pond that could be safely located some distance from the settlement to break the organics in the liquid waste. This system was suited only in towns with moderate settlement densities. In the simplified sewerage approach, which was developed in Brazil, wastewater was conveyed *via* the sewer network to a centralised but local treatment facility for the composting of the solids and the treatment of the liquid portion of the wastewater. The treatment facility was sized for the sewerage input of a neighborhood or district (e.g., 250 to 1,000 homes) as opposed to the massive scale of a centralised sewerage treatment plant for a city (e.g., serving thousands of homes and businesses). While this approach relied on a more sophisticated and centralised wastewater treatment facility than the settled sewerage approach, more natural wastewater techniques such as wetlands or aquatic treatment of the liquid portion of the wastewater could be more economically applied than complex, industrial-style treatment systems.[83]

In recent years, the development of the sewer less option showed much promise. According to Otter Wasser, "the purpose of sustainable sanitation systems is the closing of the water and nutrients cycles, taking into account that the main task of sanitation is to assure highest hygienic standards in a cost-effective, environmental sustainable way, saving both water and energy and keeping soils fertile." The guiding principle of this system is the source segregation of black water (low-diluted faeces, high nutrient content), brown water (black water without urine or yellow water), yellow water (urine from no-mix-toilets and urinals), grey water (washing, cleaning, low nutrient content), storm water runoff, (very low nutrient content),

83. Svadlenka, Robert, (2002). *Sustainable Sanitation Approaches for the Urban Poor*, *www.worldhungeryear.com*

and bio-degradable and non-bio-degradable kitchen waste. Kitchen waste and black water contained nearly all of the nitrogen and phosphorus nutrients and can be treated along with brown water through anaerobic digestion or composting to produce fertilisers and biogas. Grey water can be easily treated to a reusable quality as it had no contact with toilet wastewater. Biofilm systems like trickling filters, rotating disk or sand filters (technical or as constructed wetland) could recycle nutrients and be used for garden and washing purposes. Storm water can be used for the recharging of groundwater. If collected separately, non-biodegradable solid waste could be reduced and easily recycled. In Ottter's view, these principles were in fact well understood in developing countries and all its components—appropriate toilets, biogas plants and wastewater treatment facilities were in existence there.[84]

Experts engaged in research and application of SSWM principles have counseled both patience and caution in the matter of application to optimise benefits. Adrian Coad says, "new ideas are seldom entirely satisfactory when they are first implemented. There should be trial and evaluation before widespread implementation. After several trial and error iterations, it may be possible to introduce the new concept as a useful innovation in solid waste management."[85] Van de Klundert and Anschutz say that SSWM is a framework for change, and change inspires resistance even among people who know that the current situation is not good. Some of these resisters can be convinced through peer discussions or demonstrations. Co-opting them 'by giving them a high profile but largely symbolic role in the system' can neutralise the skeptics. In general, visual aids and live presentations are very effective, if they answer two key questions: "Where has it been done and what does it cost?"[86]

Political Importance

The Brundtland Report made an observation about sustainable development which is applicable to all its facets. It states "We do not

84. Wasser, Otter, (2002), *Sustainable Sanitation*, *www.otterwasser.de*

85. Coad, Adrian, "Introduction", in Sneha Palnitkar (ed.) *Manual on Solid Waste Management*, All India Institute of Local Self Government, Mumbai, p. 8.

86. Van de Klundert, Arnold and Justine Anschutz, *op. cit.*, p. 34.

pretend that the process is easy or straightforward. Painful choices have to be made. Thus, in the final analysis, sustainable development must rest on political will."[87] Every governmental policy is *prima facie* an expression of political will. However, when a policy is finalised in a detailed form through extended negotiations with concerned groups with a view to evolving a consensus and simple easily comprehensible steps are identified for its implementation then 'strong' rather than 'routine' political will can be said to be evident.

In some low-income countries such as India from the latter half of the 1990s there is growing evidence to suggest that local, regional and national level policy-makers and officials associated with the management of non-hazardous, human and hazardous waste are supporting the SSWM paradigm in a 'strong' rather than 'routine' way. They are willing to make a break with existing practices and supporting options which lead to reduction, reuse and recycling and hazard minimisation processes in each of these wastes. Evidence seems much less in relation to human and hazardous waste as compared to MSW for which comprehensive rules applicable to every municipality for collection, segregation, storage, transportation, processing and disposal of municipal solid wastes have been notified in 2000 under the Environment Protection Act 1986 after a tortuous process that began with the filing of a Public Interest Litigation in the Supreme Court in 1996.

The Biomedical Waste Management Handling Rules that apply to all persons that generate, handle, transport and dispose a sub-category of hazardous waste have had a trajectory involving consultation and broad review of practices comparable to the MSW Rules. In connection with human waste, it is interesting that in Kerala some municipalities have projects with CBOs and an NGO called Ecosan whose sanitation technologies have been successfully applied in 15 other countries to create toilet facilities where human waste gets segregated at source and becomes a source of fertiliser and soil conditioner and by-passes water bodies altogether. In Delhi, the MCD's successful attempt in December 2003 in organising a large group interactive event involving more than 300 persons with representatives of housewives, resident welfare associations, Eco-

87. Brundtland, G., *op. cit.*, p. 9.

clubs, students, NGOs, ragpickers and *safai karamcharis* to enable the municipality to 'know' the other stakeholders and to secure their cooperation in the task of collection of segregated waste can be regarded as a significant event in the history of SWM in Delhi. Recently, MCD has also carried out an in-house exercise for devising a payment slab on the basis of which residential colonies could be charged for door-to-door collection of waste and created a schedule for holding health camps for *safai karamcharis* who were most directly exposed to waste hazards. Similar enlightened initiatives by appropriate government agencies can be expected in the context of hazardous waste generated by industries because the High Power Committee appointed by the Supreme Court which as the final interpreter of the Constitution enjoys high legitimacy for influencing the framework of governance presented a detailed report in 2001 in which it showed how hazards to health and the environment in the existing system of hazard waste management could be minimised by incorporating the principles mentioned in Chapter 20 of Agenda 21.

A congenial environment which has both international and domestic aspects has helped in the nurture of 'strong' political will in SWM matters. At the Habitats 1I conference held at Istanbul in 1996 and at Istanbul+5 at New York, five years later, governments acknowledged the emergent trend of 'urbanisation of poverty' with three fourth of the world's urban dwellers expected to live in cities in the developing world by 2020. Mumbai, Pune and Bangalore were singled out as sources of models worth emulating because there were specific instances when the provincial government, local governments and community based organisations by acting together had been able to put in place self sustaining water supply and sanitation arrangements.[88]

Since 1992, the sustainable development paradigm, the parent paradigm of SSWM with its emphasis on the centrality of the individual, theme of timeless harmony between human beings and nature, together with its multi-disciplinary focus on economics, culture and social structures has proved to be increasingly persuasive all over the world. For many it bridged the gap between the grand ideology related alternatives of the 1950s and the 1960s and the focus

---

88. United Nations Human Settlement Programme, Istanbul+5, New York 2001, *www. Unhabitat.org/instanbul+5*.

on the individual in the 1970s. In 1999 it was said, "There is only one alternative to sustainability and that is unsustainability."[89] James Gustave Speth, a leading authority on international environment issues, said that there was a consensus in the preparatory meetings for the World Summit on Sustainable Development (WSSD) held at Johannesburg ten years after Rio that the chief issue was "implementation of the excellent agreements arrived at Rio that had failed to see the light of day due to the decline of development assistance."[90]

By the end of the WSSD, though there were no new financial commitments, it is significant that the few agreements involving targets and time tables that were made included two that had a strong SSWM content. There was an agreement to halve by 2015 the proportion of people who do not have basic sanitation. Governments that had gathered at WSSD also agreed to the aim of achieving by 2020 the minimisation of significant adverse effects on human health and the environment in the way chemicals are produced and used. Further, it is said WSSD pioneered the promotion of the track II outcomes the public-private and other partnership initiatives for sustainable development.[91] For a low income country this meant that there were addtional sources of funds from abroad for making improvements since the late-1980s in their urban infrastructure. Another supportive factor of SSWM was the emergence of new wisdom on urban governance. According to the Commission on Global Governance, "governance includes formal instructions and regimes empowered to enforce compliance as well as informal arrangements that people and institutions have agreed to and perceive to be in their interest."[92] The some perception about broad-based participation had been foregrounded in the SSWM perspective evolved in Rio.

Coming to domestic factors, the passage of the 74th amendment of Constitution in India in 1992 which required mandatory elections

89. Bell, Simon and Stephen Morse (2003), *Measuring Sustainability-Learning From Doing*, p. 1, Earthscan Publications Limited, London.

90. Speth, James Gustav (2003), "Johannesburg Summit," in *Environment*, Volume 45, Number 1, p. 27.

91. *Ibid*, p. 28.

92. Mehra, Ajay, K. (2003),"Governing India's First City," Seminar on *Sustainable Development of Delhi*, June 16-17, p. 1, Institute for Human Development, New Delhi.

to the local bodies in urban areas, was a significant event. There were provisions for additional institutions like Ward Committees and District Planning Committees when urban populations were of a certain size to enable citizens to participate in short term and long term expenditure programmes in relation to subjects listed in the 12th schedule of the Indian Constitution. SWM was explicitly made one of these. In future years, new initiatives for more durable SWM arrangements could be expected against the background of these measures.

However, if there was one factor which explained the continuing interest in SSWM in developing countries, in recent years despite its weak uneven and transient impact, it was the widely shared experiential realisation that there had been a sharp decline in environment quality. The Environment Report for Asia and Pacific 2002 corroborated this sentiment. The Report noted that successive periodic reviews since the mid-1990s of the state of the environment in regions dominated by low and middle income countries pointed towards improvements in economic growth but worsening of the condition of the environment. In the case of the Asia-Pacific region, socio-economic trends showed improvement in most sub-regions of Asia and the Pacific. Among the environmental trends, only tree plantation showed a positive trend. In the case of pollution, all indicators (freshwater and coastal pollution, air pollution, pollution by energy use, industrial pollution, vehicular pollution, green house gases, solid waste generation and agro-chemical use) showed strong negative movement for South Asia, North-East Asia and South-East Asia. The trends were also negative for Pacific and Central Asia and most are strongly so.[93] Unlike in the past, cold hard environmental facts only strengthened the ordinary citizen's anxiety for a new more durable solution in the public domain to the problem of pollution. The four year long marathon sculpting process which culminated in the notification of Municipal Solid Waste (Management and Handling) Rules in 2000 in India could not have been possible without this emotional energy.

---

93. *Environmental Governance for Sustainable Development in Asia and the Pacific* (2002), p. 39, United Nations, New York.

Conclusion

The SSWM paradigm is to SWM what the Directive Principles of State Policy is to law-making in India. The language of both clusters of principles is normative, non-sectarian and emphasises basic human needs. Protection of 'public health' in a broad sense is the goal SSWM just as Justice in its widest sense is the object of the Directive Principles. The absence of legal sanctions in the Directive Principles suggested to some Constituent Assembly members that these would prove 'useless' because they could be ignored. Others counselled that their sanction was political, and there was always the next election where neglect and disregard would have to be convincingly explained if power was to be retained. Even after 50 years, the Directive Principles continue to demonstrate their value in serving as a compass for guiding public debate and as a touchstone for evaluating the performance of politicians and beauracrats. The SSWM paradigm deserves to be better understood to serve the same purpose for waste related issues.

3

Municipal Solid Waste: Non-Hazardous Waste

Nature of Municipal Solid Waste

Municipal solid waste (MSW) or 'garbage', 'trash' and 'rubbish' as it is generally known, is generated by households, markets, commercial establishments such as hotels, institutions, offices, education centres, hospitals and industries and finally, due to the sweeping of streets. Sneha Palnitkar, Director, All India Institute of Local Self Government, Mumbai says, "solid waste is the term used to describe non-liquid waste materials arising from domestic, trade, commercial, agricultural and industrial activities and from public services."[1]

Though street sweepings may contain dead animals, faecal matter, infected biomedical waste from hospitals and nursing homes, toxic industrial residues deposited on roadsides and in vacant public land, for analytical and policy-making purposes, MSW is regarded as a non-hazardous category due to the small amount of such waste, in comparison with the totality of MSW which consists largely of food and garden related residues and non-biodegradable packaging material and construction debris.

The National Environmental Engineering Research Institution, Nagpur (NEERI) noted in 1995 in its Strategy Paper on Solid Waste Management in India that its survey of 43 cities with population ranging from 0.1 to 5 million and above revealed that Indian cities and towns generated on an average 200-500 grams of municipal solid waste per person per day. The rates of generation were low in small towns and high in big cities.

The physical characteristics of MSW in Indian cities consisted of: paper, (2.91 per cent to 6.43 per cent); rubber, leather and synthetics

1. Palnitkar, Sneha, "Municipal Solid Wastes," in Sneha Palnitkar (ed.) *Manual on Solid Waste Management*, p. 9, All India Institute of Local Self Government, Mumbai.

(0.28 per cent to 0.78 per cent); glass (0.46 per cent to 0.94 per cent); metal (0.32 per cent to 0.80 per cent); total compostable matter (30.84 per cent to 56.67 per cent); and inert material (40.07 per cent to 53.90 per cent). The chemical characteristics of MSW were found to be; nitrogen (0.56 per cent to 0.71 per cent), phosphorus (0.52 per cent to 0.83 per cent) and potassium (0.52 per cent to 0.83 per cent) with C/N ratio between 21 and 31, calorific values in the range between 800 and 1010 of K.cal/kg. and the density of waste between 330 and 560 kg/cubic meter.

The NEERI Report concluded that in the Indian waste there was a small percentage of recyclable material and more of compostable and inert materials.[2] Inert waste, such as ash and road dust, comprises material that does not give rise to harmful chemical or biological agents when placed in a landfill.

In the case of municipal solid waste in Delhi, the components were identified in 1994 as bio-degradable 38 per cent, inert 35 per cent, bio-resistant 14 per cent, plastic 6 per cent, paper 6 per cent and glass and crockery 1 per cent. (Figure 3.1).[3]

Figure 3.1

Composition of Municipal Solid Waste (MSW) in Delhi

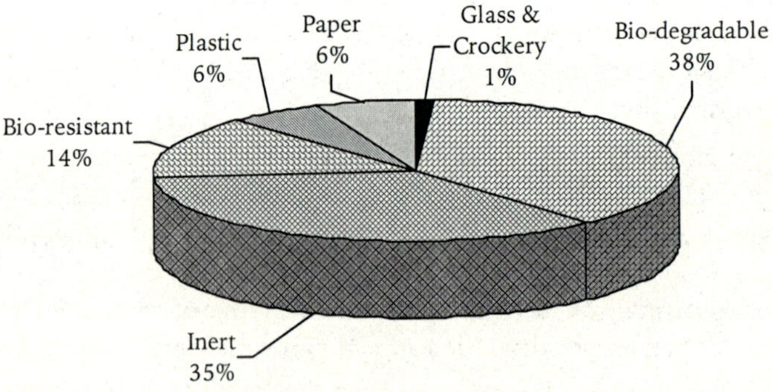

2. National Environment Engineering Research Institute (NEERI), (1995). *Strategy Paper on SWM in India*, August, Nagpur.
3. The Energy Research Institute (TERI), *State of Environment Report for Delhi 2001*, p. 67, New Delhi.

The quantity of waste generated has steadily grown in recent years. As per a NEERI estimate, cited in the State of Environment Report for Delhi 2001, the present solid waste generation in the National Capital Territory (NCT) of Delhi is 6,000-7,000 tonnes per day.[4] The Conservancy and Sanitation Engineering Department of the MCD, which is responsible for the management of solid waste and storm water drainage system in about 94 per cent area of Delhi, considers the average garbage generation as 6,000-6,500 tonnes per day or the per capita generation as 450 grams.[5] The waste generation in Delhi is slightly less than the waste produced in Mumbai which is 6,640.20 tonnes per day but much more than waste produced by other mega cities in India such as Ahmedabad (2,087 tonnes), Bangalore (2,480 tonnes), Chennai (2,087 tonnes), and Kolkata (4,578 tonnes).[6] At a growth rate of 6 to 8 per cent (economic growth plus two per cent), quantity of waste produced is likely to be between 17,000 and 25,000 tonnes per day in 2021.[7]

Low-income communities in Delhi, like in other Indian cities, produce less waste as compared to high-income communities. A joint study carried out in Delhi by the Tata Environment Research Institute (TERI) and the NGO, Srishti, in 1996 noted that in slum and resettlement colony clusters where over 40 per cent of Delhi's population lives, the waste generation was 200 grams per day in respect of households having monthly income of Rs. 2,000 while it was 800 grams per day in more affluent localities containing households having a monthly income of Rs. 8,000 and above. Not only did the quantity vary, there was also a difference in the composition of the waste generated. The more affluent localities generated more organic waste than the low-income localities. The more affluent localities also generated more paper and plastic waste while the low-income localities generated more grit and glass waste.[8] (Table 3.1)

---

4. *Ibid*, p. 65.
5. Municipal Corporation of Delhi, *Background Note on Conservancy and Sanitation Engineering Department*, p. 1.
6. Srishti, (2002), "Report on Recycling Responsibility, Traditional Systems and New Challenges of Urban Waste Management in India," p. 14, June, New Delhi.
7. Delhi Urban Environment and Infrastructure Improvement Project (DUEIIP) Report, (2001), *Delhi-21*, Government of India, and Government of National Capital Territory of Delhi, p. 49, January.
8. Srishti, *op. cit.*, pp. 14-15.

Table 3.1

Variation in Types of Waste Generated in Low-income and High-income Communities in Delhi

(Per cent)

Nature of Waste	*Low-Income Households*	*High-Income Households*
Organic (wet)	72	82
Grit	14	*
Plastic	5	8
Paper	5	9
Glass	4	1

Note: * denotes Negligible.

The difference in the nature of the waste generated by more and less affluent localities in Delhi finds a parallel in the nature of waste generated in the high-income countries and low-income countries. In the case of high-income countries and affluent localities, while the component of grit in solid waste is negligible, the proportion of plastic and paper waste is higher in both richer colonies and countries as compared to their poorer counterparts.[9]

Seasons also impact on the nature of waste generation. Though in the festival season waste is more visible, in the MCD calendar, it is in June that garbage generation and collection is at its peak because of the pre-monsoon silting of drains and *nallahs*.[10]

Not only massive quantities of waste are generated in Delhi, the sources of waste generation are also huge in number. In the MCD area, there are 1,600 organised markets, 24,600 wholesale establishments, 100 weekly markets, 6,000 make shift shopping spaces and 1,40,000 informal retail units. In addition, there are more than 80 hospitals and 1,000 nursing homes. Besides there are wholesale fruits and vegetable markets, truck terminals and slaughterhouses, "with their filth, squalor and unsanitary conditions."[11]

9. Cointreau-Levine, Sandra (1982). *Environmental Management of Urban Solid Wastes in Developing Countries*, A Project Guide for the World Bank Staff, p. xii.

10. Ghosh, Archana (2000). *Solid Waste Management in Delhi, An Exploratory Study on Local Government-Community Interface*, p. 32, Institute of Social Science, New Delhi.

11. Interim Report of the *First Finance Commission for NCT of Delhi* 1996, quoted in The Energy Research Institute's Final Report (draft) No. 1999EE41 (March 2002), Urban Services Environmental Rating System, p. 169.

Most of the waste generated by industries is directly picked up by recycling units and bulky construction and demolition waste is disposed of by owners by payment to private contractors for dumping into low lying areas. Thus the municipality has the direct responsibility for collection, safe disposal of a large fraction of the total solid waste that is routinely generated. 56 per cent of the total waste collected was from the residential areas. These include both wastes from private premises and through sweeping operations on the streets.[12]

Problem of Solid Waste Management in Delhi

An enduring view about Delhi is that India's first city is also among its dirtiest. While "garbage does leave people's homes, it never really leaves one's sight."[13] Solid waste management—the collection, transportation, treatment, composting and final disposal of waste—from streets and dustbins not from individual homes or institutions is meant to be carried out in 94 per cent of Delhi's territory by the Municipal Corporation of Delhi (MCD). It discharges these obligations/functions, which are indicated in the Delhi Municipal Amendment Act, 1993, after a fashion even in relation to the up market residential localities. There may not be heaps of garbage by the roadside in these areas but the common spectacle of public dustbins, with as much trash outside as inside, is symbolic of how unsatisfactory the situation is in most parts of the city.

Since the late 1980s, a number of community based initiatives led by NGOs have tried to ameliorate this condition for which a variety of reasons concerning the municipality such as paucity of funds, growing costs of services, and indiscipline in the work force are put forward. Some NGOs such as the Centre of Science and Environment have focussed mainly on awareness building about the adverse environment and health impacts of present day practices and advocacy of alternatives. Others such as Vatavaran and Srishti now know as Toxics Link have worked with resident associations in established

12. The Energy Research Institute's Final Report (draft) No. 1999EE41 (March 2002), Urban Services Environmental Rating System, p. 170.

13. Venkateswaran, Sandhya (1995). *Alternative Solid Waste Management Systems, Eight Case Studies from Across the Country*, National Institute of Urban Affairs, New Delhi, p. 24.

residential colonies to collect and reduce waste through segregation, resource recovery and composting. Vatavaran also runs projects for exchange of recyclable waste with organic manure and recycled paper products at the individual household level. Still others such as the Asian Centre for Organisation Research and Development (ACORD) have nurtured collective initiatives in slums and demonstrated how a sustainable system can be set up where waste from being a nuisance becomes a domestic resource and an income supplement.

Given the size of the city, the impact of these endeavours has been limited because the primary stakeholder, the MCD, unlike its counterparts in other cities such as Bangalore, Chennai and Hyderabad has been indifferent in the matter of multiplying or up scaling such scattered projects.[14]

Since the mid-1990s, the MCD's performance has been a subject of intense public scrutiny. In 1994, the Supreme Court while admitting the writ petition filed by Dr. B.L. Wadhera, seeking directions to the MCD and the NDMC to compel them to perform their statutory duties in the collection, removal and disposal of garbage and other wastes in the city efficiently, asserted that neither the non-availability of funds nor inefficiency of staff could absolve these bodies from their responsibilities.

The Supreme Court issued a 14 point directive not only to the MCD and the NDMC, but also to the concerned departments of the Central government and the government of NCT Delhi and ordered them to submit a compliance report separately to the Court. Some of the significant directives addressed to the MCD and the NDMC were:

- Experiment with door-to-door collection of garbage by distributing polythene bags and their daily clearing.
- Clean and scavenge the city and carry garbage from the collection centres to the designated sites every day.
- Ensure cleanliness and maintenance of receptacles daily.
- Ensure that garbage is not found spread around the collection centres and on the streets.
- Activate the defunct compost plant in Okhla and explore the possibility of constructing four additional plants.

14. Ghosh, Archana (2000). *op. cit.*, p. 57.

- Open direct communication lines with people to educate them about garbage disposal.[15]

In the wake of Supreme Court's orders, the MCD appointed NEERI to make suggestions with cost implications on how it could improve its functioning. After making an in-depth analysis of the situation, NEERI in its Report in 1996 recommended specific technological and managerial innovations and appropriate financial provisions for all stages of solid waste management operations: collection, transportation, treatment and final disposal to be implemented within the short term. The Report generally emphasised the forging of new links between the Municipality, the private sector, NGOs and citizen groups for adaptation of technology to local needs, improvement of work norms and supervision, strengthening of infrastructure, and capacity building at lower levels. These changes were all in line with the suggestions of the Sustainable Solid Waste Management paradigm.

NEERI in its Report recommended sharing by the MCD of the task of collection and transportation of MSW in select areas with private operators. Private companies could be hired to collect and transport waste from sources such as hotels, markets and housing societies. It further suggested mobilisation of NGOs and the use of print and electronic media to generate awareness of danger to health and the environment from current waste management practices and the need of promotion of waste segregation at source. About waste collection centres, it recommended that these should have roofs, be well lit and be better designed to facilitate complete removal of waste. A *Safai Karamchari* (cleaning staff) should be posted at these points or *dhalaos* round the clock to prevent spreading of the collected waste by 'ragpickers and stray animals.' Community bins of 50 litre capacity should be placed 50 m apart and open top bins replaced with closed bins to prevent soaking of the waste by rain water during the monsoons.

With a view to replace the time consuming manual method of loading of the vehicles, the NEERI Report recommended that transport vehicles should include more front-end loaders for mechanical lifting of solid waste. Dumper placer-containers were needed to serve as

---

15. *Ibid*, p. 7.

community bins in suitable locations. Vehicles that were more than seven years old were required to be replaced with new vehicles with closed bodies specially designed for solid waste. Existing workshop facilities were identified for renovation and augmentation for improved maintenance of vehicles.

The report also emphasised that the existing work force of *Safai Karamcharis* (SKs) and supervisory level staff would have to be deployed more efficiently. Supervisory level staff would also have to be strengthened in numbers so as to cover poorly serviced areas within the jurisdiction of the MCD and enhance control over waste at landfill sites. Not only were resettlement colonies and JJ clusters inadequately provided for, the 1,300 unauthorised colonies and the PWD roads were not cleaned and swept at all. The SKs needed to be provided with better quality equipment such as brooms and baskets protective gear such as gumboots, gloves, and headgear secure space for storage of work materials; and comprehensive health insurance. New work norms needed to be negotiated with the SKs and their compliance ensured. Though the Sanitary/Assistant Sanitary Inspectors supervised the work of the SKs, there was no formal system of reporting. A map prepared on GIS depicting different features of SWM system, like *dhalaos*, roads, offices, landfills etc. could be an invaluable tool for preparing fresh plans for deployment of vehicles and manpower and for their monitoring.

NEERI Report further said that the MCD's management of landfill was the most neglected aspect of its SWM responsibilities. There were no fences to prevent entry of unauthorised persons in the landfills such as of rag-pickers, no roads within the site resulting in difficulties in movement of vehicles, no system for covering of incoming waste or for composting the non-bio-degradable component and for detection and recovery of gas which resulted in frequent fires. A mixture of flyash, soil and stabilised material was identified as appropriate covering material whose application needed to begin without further delay. In addition, a system for leachate collection and treatment, equipment for fire fighting and increase in the number of bulldozers for compacting waste was indicated. Taking the view that waste reduction was an important aspect of waste management, NEERI suggested constructions of additional compost manufacturing plants and cooperation with the Horticulture Department in utilising compost. Funds required for implementing the various initiatives were also indicated.[16]

Between 1996 and 1998, MCD's Conservancy and Sanitation Engineering Department officials constructed a few models of more user-friendly *dhalaos* but further experimentation was abandoned due to differences over cost benefits. Door-to-door collection of waste, rather than from merely the public *dhalaos* as required by the MCD Act, 1957, through trucks which declared their arrival at pre-designated places by musical horns was attempted in some localities. Segregated waste collection in previously distributed black and red bags was also tried out in specific colonies. Both programmes were abandoned because neither were the trucks getting the minimal load expected and nor were the bags returned with adequate quantities of waste, if returned at all. The MCD also launched periodic cleanliness drives. The number of *dhalaos* or masonry structures with roofs went up from 412 to 798 and dustbins or structures without roofs, from 1,184 to 1,428. As per the MCD records, the collection of garbage between 1996 and 1998 increased from 3550.49 tonnes a day to 5,000 tonnes a day.[17]

However, the Central Pollution Control Board (CPCB) which had been directed by the Supreme Court to report upon the progress of the implementation of its orders, and which had prepared 12 reports on the subject by January 1998, was not impressed by MCD's efforts. In its first report to the High Court in March 1998 (the Supreme Court had transferred this authority to the Delhi High Court in January 1998), it noted that the MCD waste collection centres continued to be dilapidated and overflowing with garbage. Transportation of the waste to the landfills, besides being carried out in an unhygienic manner was not being done every day. Waste reduction through removal of recyclables was only haphazardly done because it was dependent on rag-pickers rummaging at dump-sites. Further, only 37 tonnes from the 5,500 tonnes of solid waste that the MCD collected every day was reaching the Okhla compost plant. The MCD needed to improve its waste management operations and

---

16. National Environment Engineering Research Institute (NEERI) (1996). "Action Plan for Immediate Improvement, Nagpur," *Solid Waste Management in MCD Area, Phase 1 Report.*

17. Ghosh, Archana, *op. cit.*, pp. 28, 32, 48, 49.

mobilise people's cooperation to discharge its statutory responsibilities adequately. "Lack of proper communication between the public and the MCD staff and between different categories of municipal staff culminates into delayed redressal of public grievances and public participation."[18]

The CPCB also noted that the capacity of the three sanitary landfills at Gazipur (area 70 acre), Bhalswa (area 40 acre) and Okhla (area 18 acre) was due to reach saturation point in a few years and MCD was likely to face a huge crisis soon unless it became pro-active in identifying and finalising additional landfill spaces.

In CPCB's view, neither the MCD nor the government at Delhi or at the Centre displayed any sense of urgency to introduce wide ranging manpower and technology related reforms, provision of adequate equipment and civil and electrical works at landfill sites for which NEERI had suggested an allocation of over Rs. 72 crore.[19] As late as 2000, only fencing had been provided at the landfill sites to prevent unauthorised entry and application begun in a haphazard manner of a 15 cm thick layer of construction wastes/earth in lieu of 0.5 meter thick layer of fly ash, soil and stabilised material that had been proposed.[20]

It is apparent that between 1996 and 1998, the MCD did not get much help from the Delhi government or the Central government to implement the recommendations of NEERI.

Social and Environmental Implications

The presence of an informal sector in Delhi, as is the case in other Indian cities, has served to mitigate the adverse impacts on health and the environment arising out of the unsatisfactory discharge of duties by the MCD. The informal sector also provides opportunities for employment for low-income groups and resource recovery. However, these goals are achieved at great cost to individual/public health and the environment due to the constraints within which it functions.

---

18. *Ibid*, pp. 7-8.
19. NEERI (1996), *op. cit.*
20. Ghosh, Archana, *op. cit.*, p. 36.

A heap of garbage, which is a source of stench besides being an eyesore, serves as a workstation for the lowest level of a hierarchy that supplements the municipality's effort to collect and finally dispose of solid waste. In Delhi, 10 per cent to 15 per cent of waste is taken care of by it. This hierarchy, which at its higher level is increasingly organised, at its base, consists of waste-pickers, more commonly known in India as 'ragpickers.' In addition, there are waste dealers or *kabadiwallahs,* agents and finally reprocessors or recyclers. Waste dealers can be classified as small waste dealers who generally buy from the waste-pickers and buy all types of recyclable items and medium waste dealers who buy only select items and large waste dealers-cum-suppliers who generally buy from both the small and medium waste dealers and may specialise in just one or two items. Agents facilitate transactions between the medium and large waste dealers and the recyclers.[21]

Waste in order to be recycled requires very precise segregation. Plastics and paper, in particular, can be of various types and are supplied to the reprocessors segregated according to their reuse and chemical composition determined by physical appearance and texture, a method which is quickly grasped by an entrant into the informal waste management sector. The segregation and classification of waste continues throughout most of the recycling chain till the waste, duly segregated, reaches industries that process it and generate feedstock for recycling industries. There is "close inter-dependence in many ways of the layers—from top to bottom—upon each other." It has been estimated that more than 1,000 tonnes of recyclables are recovered from the waste which can undergo value addition of upto 700 per cent and contributes to the feedstock of over 1,000 recycling units.[22]

Delhi has approximately 90,000 to 100,000 waste-pickers. In a recent survey carried out by the NGO, Srishti, in different parts of Delhi from June 2001 to January 2002, it was found that among waste pickers, 24 per cent were women, 24 per cent were under the age of 16 and 77 per cent were illiterate. In all probability, as the study suggests, the number of women and children who engage in this

---

21. Srishti, *Report on Recycling Responsibility, op. cit.,* p. 44.

22. Chaturvedi, Bharati (September 2003). "Waste-handlers and Recycling in Urban India: Policy, Perception and the Law," *Social Change,* Vol. 33 Nos. 2 and 3, June, pp. 42-43 and Srishti, *Report on Recycling Responsibility, op. cit.,* p. 44.

activity is much more than what its survey based on a small sample size found. The regional and cultural backgrounds are a major factor that affects the participation of women and children in this trade. Most of the Muslim families did not allow their women and adolescent girls to engage in the waste picking activity.[23] An earlier study of a slum in West Delhi where mainly waste picking families from Bihar and Tamil Nadu lived, had found that while among the Tamil community, entire families were engaged in waste picking, the Bihari community was categorical about not letting its women and girls engage in this work.[24]

The Srishti study, like earlier studies about migrant population, found that 97.5 per cent of the waste-pickers were migrants, with 78 per cent from four states—West Bengal (33 per cent), Uttar Pradesh (22 per cent), Bihar (13 per cent), and Bangladesh (10 per cent). Their stated reason for coming into the city was either poverty or an ecological disaster. Since the majority of them were illiterate, waste picking was the only available option for subsistence which perforce had to be engaged in on every day.[25]

In sifting through garbage, waste-pickers could get bitten by rodents because rodents are attracted by unattended garbage; and they could be injured as broken light bulbs and broken glass bottles and biomedical waste such as a used syringe could well be a part of the waste. According to the Srishti study, 28 per cent of waste-pickers reported having such injuries often while 61 per cent had experienced them sometime. In the case of handlers of biomedical waste, 27 per cent of the waste-pickers reported injuries from syringes and sharps.[26] The waste-pickers could also be infected by pathogens which begin breeding in the waste. Special epidemiological studies have shown that workers engaged in SWM services frequently suffer from respiratory tract infections, gastro-intestinal parasites, skin disorders and eye infections.[27]

---

23. Srishti, *Report on Recycling Responsibility, op. cit.*, pp. 32-37.
24. Venketeswaran, Sandhya, quoted in Srishti, "Report on Recycling Responsibility," *op. cit.*, p. 36.
25. Srishti, *Report on Recycling Responsibility, op. cit.*, p. 37.
26. *Ibid*, p. 41
27. Report of the Committee constituted by the Hon. Supreme Court of India, *Solid Waste Management in Class 1 Cities in India*, March '99, p. 78.

It is not the waste-picker alone whose health is at risk due to uncollected garbage. Pathogens become airborne and transmit diseases like malaria to people living in the vicinity. Surface run offs can contaminate water bodies and groundwater. Strewn garbage creates drain blockages and stagnant pools of wastewater in which disease vectors flourish. Neglected piles of accumulated garbage lead to blocked sewers and wastewater backflows that can flood homes.

The slum-dwellings where the waste-pickers live also serve as storehouses of waste due to their dependency relationship with the more prosperous middle level dealer who could be the actual owner of the residential space. A major component of waste is plastic waste, some varieties of which such as polystyrene, used for hot beverage cups, food trays and egg cartons are highly inflammable. According to the records of Delhi's Fire department, "almost 75 per cent of the fires in slums are due to stored plastics" and "from 1989 to 1998 approximately 69,147 *Jhuggis* were badly affected by such fires." Most deaths in these fires were due to "inhalation of toxic gases which are liberated when plastics/PVC are involved." In recent years, one of the biggest fires in Delhi occurred in 1995 in Jawalapuri Plastic Market in which about "4,000 godowns, 50,000 workers and 2,500 other sorters" were affected. Huge quantities of PVC, the most retrieved form of plastic, greatly coveted because of its high market value since it is the ideal material for a variety of consumer products such as mugs, slippers and hangers, were burnt up in this fire.[28]

According to the Srishti study, a waste worker earned Rs. 45 to Rs. 80 a day whereas a child could make up to Rs. 30. A waste-picker who owned a bicycle could earn between Rs. 50 and Rs. 80 and a tricycle owner could earn up to Rs. 200. However, these figures are predicated on the continuity of the recycling industry which actually rests on a very shaky ground since it is, by and large, in the informal sector. In Delhi, thousands of such units till recently were located in non-conforming areas and 2001 was a tumultuous year not only for workers but also suppliers since the Delhi government in pursuance of the Supreme Court's orders decided to withhold the annual renewal of licences, which made their existence legal.

---

28. Chaturvedi, Bharati (2000). "Plastic Not So Fantastic," *Hindustan Times*, September 9, Delhi Edition.

The recycling industry owner's limited capital prevents him from optimising production or modifying his machines to meet the requirement of new packaging materials such as fused aluminium and plastic sheets. This in turn inhibits a waste-picker from picking packaging waste of the kind used for keeping potato chips fresh. The recycler also tends to use devices that are fabricated from second hand or discarded machines with a minimum of parts and near absence of safety features which puts the worker at great risk and who, like the waste worker, has little choice in the matter of employment. Workers engaged in melting aluminium cans, a favoured item because even with repeated recycling there is no decline in its intrinsic value, are prone to burn injuries. Aluminium dust is strongly fibrogenic. Aluminium dust can cause nodular lung fibrosis and interstitial lung fibrosis and emphysema as indicated in experiments with animals.

During the Srishti survey, it was found the PVC used for making soles for chappals was being recycled at Narela and Mangolpuri. PVC recycling involves re-melting and then re-forming the used plastics. In Jawalpuri, Seemapuri and some other places, PVC coated copper and aluminium electric wires are burnt to recover the metal. Burning of PVC emits dioxins. Dioxins and dioxin like compounds are one of the most toxic chemicals in the world. They have an extremely long life in the environment. Dioxins are transported through the atmosphere and eventually deposit on soil, plants and water bodies and because they are fat-soluble get passed on to people through food items such as meat, milk and fatty food products. Tiny doses of dioxins can cause cancer, endocrinal disruption, immune system depression and problems in the reproductive systems.[29]

The recycling process is not only hazardous for the workers but also for the community at large. Since the recyclers do not follow any standards such as 'prime grade,' 'once recycled,' 'twice recycled,' a recycled product improperly labelled as 'food grade,' could cause great harm to a buyer of the product. Shymala Mani has noted that to hide the dullness in the colour of recycled plastic products, "Non-permitted colours and additives are used in excess leading to poor quality of products thereby rendering them unsafe and dangerous." She has further observed that excess quantity of sub-standard chlorinated

29. Srishti, *Report on Recycling Responsibility, op. cit,* pp. 42-43,47,77.

paraffin wax (CPW) is added to reprocessed PVC cables to make them shiny and new looking. In summer when the temperature reaches about 50 degree Celsius, this substance starts melting from these cables creating gaps which causes short circuits and fires. "Extremely poor quality PVC cables with high quantities of plasticisers such as CPW are responsible for the large number of fires that have now become common in tall buildings (e.g. Nehru Place), marriage halls and even during functions with temporary electrical connections in Delhi and many parts of India."[30]

The failure to take steps to neutralise and mitigate impact of leachate formation at the landfill sites has meant that heavy metals such as lead, mercury and cyanides can seep into the groundwater and contaminate it leading to worsening of the drinking water resource position since Delhi relies upon it heavily to augment the supply from the Yamuna. NEERI had suggested that fire-fighting equipment and water sprinklers be provided at the landfill sites and pipes put in place to recover methane and utilise it for producing electricity. No action was taken in relation to these recommendations until 1998.

Government of India Initiatives

In December 1998 when the Congress party won the state level elections and formed the government in Delhi, a comprehensive framework requiring radical changes in the manner in which solid waste was until then collected, transported and finally disposed of by local authorities was on the threshold of becoming the law of the land. The Central government had issued a draft notification on the management of municipal solid waste which was made available to the public in September 1999. After considering the objections and suggestions received in respect of the Rules, the Government of India, in exercise of the powers conferred by Sections 3, 6 and 25 of the Environment (Protection) Act, 1986, notified the Municipal Solid Waste (Management and Handling) Rules, 2000 in 2002.[31]

The notification of the MSW Rules was not a sudden event and nor were the contents of the Rules a matter of surprise to municipal

---

30. Mani, Shyamala K., *et al.*, *Less Plastics for a Better Future*, Centre for Environment Education, Delhi, p. 11.

31. *Municipal Solid Waste (Management and Handling) Rules, 2000.*

and state level governmental authorities. The outbreak of suspected plague like disease in Surat in 1994, which had caught media attention the world over, had impressed upon the Central government that practically all Indian cities were equally vulnerable and the need to improve their condition should be a top priority. The Government of India had long recognised the futility of the passive policy, followed since the 1960s and 1970s, of managing municipal waste by offering loans and grants to urban local bodies to set up composting facilities. The National Waste Management Council set up by the Central government in 1990 seemed to be interminably pondering over the problem of recyclable waste posed by 22 municipalities.[32]

The High Power Committee on Urban Solid Waste Management in India, set up by the Planning Commission in 1994, under the Chairmanship of J.L. Bajaj, Member, Planning Commission with 11 other experts in the wake of the suspected plague in Surat, had submitted its Report in May 1995. Its assessment was that, "a situation analysis of the existing state of waste management in the Indian cities and towns would lead to the inevitable conclusion that better sanitation standards could have been achieved in most of our cities and towns by prudent and planned allocation of available resources to develop and support the application of appropriate low cost, eco-friendly technologies." The committee called for the formulation of a National Policy on Solid Waste Management but since the issue had urgency it also suggested an action plan for the last two years of the Eighth Five Year Plan for all cities.

The Bajaj Committee, focussing on reduction and safe disposal of MSW understood broadly as, "urban waste including hazardous industrial and hospital wastes", suggested an immediate review of current Municipal bye-laws and necessary modifications "taking into account local regional variations for this purpose." At the State government level, the Report suggested creation in the concerned Ministry of a cell that could coordinate and oversee all aspects of solid waste management activities such as data collection, "assess and propose projects for safe management of solid waste by various municipalities, propose and obtain necessary financial outlays" and

32. Gulaty, Sachin and Santhosh Ragavan (2003). "Solid Waste Management in India," Background Paper for Workshop, *The Need for Institutionalisation of Solid Waste Management in India*, p. 6, April 26, The Energy Research Institute, New Delhi.

legislative authorisations and human resource development requirements. A network of regional centres was needed to take up research and development studies related to safe management of urban solid waste and "undertake pilot studies of economically viable procedures for safe disposal."

The Report further said that "recycling should get due recognition and support as a method of converting waste into useful articles; recycling industry should get assistance for technological upgradation to improve the quality of the product, reduce cost and minimise potential health hazards." There should be, the Report also stated, "a cooperative for ragpickers so that the middle-men are eliminated." Finally, it suggested that the stakeholders in safe management of waste, the industrial managers, health-care personnel, and the general public needed to be made more aware of their respective roles in ensuring safe disposal of potentially hazardous waste.[33]

In 1998, barely four years after, the Government of India was pulled up for neglecting the build up of waste in Delhi, it was again in the dock. It was required to identify specific measures which Central, State and Urban local bodies (ULBs) could undertake to resolve the pan Indian problem of slovenly cities. The Supreme Court had admitted the Public Interest Litigation (Writ Petition Number 888 of 1996) of Almitra H. Patel and Another *versus* the Union of India and Others, seeking directives from the apex Court to the three levels of governance to improve the Solid Waste Management practices. After a few hearings, the Supreme Court constituted a Committee of eight members under the Chairmanship of Asim Barman, Municipal Commissioner, Calcutta Municipal Corporation which included S.R. Rao, former Municipal Commissioner of Surat, who was widely credited with completely transforming Surat during his tenure, and P.U. Asnani, SWM technology expert who had been associated in the 1980s with innovative World Bank assisted practices in enhancing waste collection in Ahmedabad together with social activist Almitra H. Patel to look into all aspects of urban solid waste management in Class-I Cities (cities having a population of one lakh and above). The Committee was directed in particular to suggest methods of hygienic

33. Planning Commission (1995). *Urban Solid Waste Management in India*, Report of the High Powered Committee, Government of India, New Delhi, *www.planningcommission.nic.in/reports*.

processing of waste and suggest ways to improve conditions in the formal and informal sector for promoting eco-friendly sorting, collection, transportation, disposal, recycling and reuse of waste. The Committee was directed to report to the Supreme Court.[34]

In March 1999, the Barman Committee published its Report that had relevance for more than 65 per cent of India's population that was living in the 300 Class I cities. Four workshops were conducted in Calcutta, Chennai, Mumbai and New Delhi on the Interim Report that was made available in June 1998. Mayors, Municipal Commissioners, Chief Executive Officers, Heads of Departments of Urban Development/Solid Waste Management of various States, Secretaries to Government, representatives of many national and international organisations, NGOs associated with SWM, representatives of Medical Practitioners' Associations, Traders' Associations and private sector service providers and interested individuals were invited to participate in the workshops.

The Final Report reflected the end result of these deliberations as also consultations with the Central Pollution Control Board in respect to the minimum standards that local bodies should and could achieve. The Report was circulated among the State governments and despatched to concerned cities to elicit their views about implementation in the context of their institutional capacities, technological options and financial constraints. On the basis of the Report, the Supreme Court directed the Ministry of Environment and Forests of the Central government, to frame rules for the management and handling of municipal waste. The Supreme Court also ordered the five mega cities of Delhi, Bangalore, Calcutta, Chennai, and Mumbai to start implementing the Report and the CPCB was directed to monitor the process.[35]

What was most significant was that the resultant Municipal Solid Waste (Management and Handling), Rules, 2000, incorporated very substantially the sustainable solid waste management principles that had been articulated at Rio, and affirmed in the Reports of the High Power Committee of the Planning Commission on Urban Solid Waste

34. *Solid Waste Management in Class 1 Cities in India, op. cit.*, p. 9.

35. Asnani, P.U., "Modernisation of Solid Waste Management Practices in India with NGO," Public and Private Sector Participation in A.H. Sinha, Md. Maqsood and Iftekhar Enayetullah, (eds.), *op. cit.*, p. 200.

Management in 1995 and more recently by the Barman Committee in 1999. Also, what was especially important in the context of Delhi was the fact that the NEERI Report's detailed recommendations were endorsed by the Rules.

The actions mandated by the Municipal Solid Waste (Management and Handling) Rules, 2000, in the context of collection, transportation and disposal of municipal solid waste are essentially those suggested by the Barman Committee to make Indian cities "clean and liveable." To understand the rationale and significance of the Rules, these need to be read in conjunction with the Barman Committee's Report.

The omissions in the succinctly formulated Rules have generated disappointment among environment NGOs. "The new Rules do not address mechanisms, which will be needed for promoting recycling or waste minimisation. Secondly, there is no provision for any public participation despite the fact that the legislation of the Rules have been an outcome of public pressure and the work done by the NGOs and the community groups in this area."[36]

This assessment was understandable considering the way the NGOs were marginalised in 2002 when the dominant municipal organisation in Delhi, the MCD proposed handing over of collection, transportation and disposal of waste to private companies in six zones in the first phase. In retrospect, the MCD should have initiated a debate with Delhi's public about the practical difficultes in implementing the Rules. Schedule I of the Rules that regard NGOs associated with solid waste management activities as virtually indispensable for operationalising the Rules.

Both the Schedule II of the Rules and the Executive Summary of the Barman Committee Report begin by stating that urban spaces should be protected from pollution. Schedule II of the Rules stated that, "littering of municipal solid waste shall be prohibited in cities," and the Barman Committee had recommended, "no waste shall be thrown on the streets, footpaths, open spaces, open drains or water bodies." The Report goes on to say that at the point of source generation, storage should be done separately for biodegradable waste and recyclable waste in a bin and a bag. Domestic hazardous waste as

36. Agarwal, Ravi and Sanjay K. Gupta, "Rethinking Urban Waste in India: Creating a Community Paradigm," *Social Change*, *op.cit.*, p. 58.

listed in its Annexure E, is specifically excluded. The Rules skip the point on separate storage and state that the municipality should ensure house-to-house collection through a stated methods such as a vehicle visit on pre-informed timings.

The Rules go on to state that waste collection method from slum squatter areas/hotels, restaurants, office and commercial complexes should be finalised by the municipality through consultations with these entities. The Rules clarify that large-scale generators of bio-degradable waste such as fruit markets and slaughterhouses require negotiated arrangements. Horticulture waste and construction and demolition waste should also be separately collected and disposed of. Biomedical and hazardous waste needed to be sanitised first and separately disposed of as per the existing MoEF Rules. It is explicitly stated, that "garbage leaves shall not be burnt."

The Report says that collection of the bio-degradable and recyclable waste could be staggered. While containerised vehicles should collect the bio-degradable portion of the waste, rag-pickers organised by the NGOs, could periodically collect the latter. On the mode of collection, the Rules prescribe that municipal authorities can determine the modalities most appropriate for "public benefit in a city or town." However, the Rules add that the municipal authorities "shall undertake a phased programme to ensure that the community is involved in waste segregation." The Report says that the streets and public places should be swept daily by rotating the rest days for the sweeping staff. The Rules require only the public waste storage sites to be cleaned daily. While the Report recommended that waste storage sites should be closed spaces, the requirement of the Rules is less stringent—there should be protective covers and the storage sites should be user-friendly. Both the Report and the Rules reject transportation in open vehicles and emphasise minimisation of manual loading and multiple handling of waste before its final disposal. About the processing of the waste, the Report is categorical. It should be maximised and this could be done in two ways; by composting the food waste and bio-degradable waste and passing on the recyclable waste to the recycling industry. The Rules stipulate that the, "Bio-degradable waste shall be processed by composting, vermicomposting, anaerobic digestion or any other appropriate bio-logical process for stabilisation of waste." The Rules further states that concentration limits of metals such as Arsenic, Cadmium, Chromium, Copper, Lead, Mercury, Nickel and Zinc should not exceed the prescribed concentration limits as the final product otherwise could not be used for growing food crops.

Clearly, the framers of the Rules desired a complete break from the existing practice of composting un-segregated municipal waste that invariably has a high heavy metal content (Table 3.2). As Shyamala Mani has pointed out there is danger from toxic elements in unsegregated waste to both human beings and animals because heavy metals can be biomagnified up the food chain. Heavy metals can also contaminate ground water aquifers.[37]

Table 3.2

Heavy Metal Concentrations in mg/100 gm in Compost Prepared from Segregated and Unsegregated Wastes

Heavy Metal	*Segregated*	*Unsegregated*
Total chromium	11.33	72.072
Lead	1.08	3.667
Cadmium	0.39	1.87
Nickel	Nil	3.958
Arsenic	Nil	Traces

As compost from waste as per new standards cannot be made unless households, NGOs and ragpickers work in close association, the Report had urged that, "decentralised composting with public and NGO/CBO participation may be encouraged wherever possible and centralised composting of the rest of the waste may be done."

The crucial importance of multi-site composting is explained in the Report in the terms of both short term and long term benefits. There was a dearth of land for setting up of waste processing plants for final disposal of waste in Indian cities and "food and bio-degradable component useful to agriculture was going waste." The Report further said, "Measures for conservation of land and organic waste resources shall be taken and organics returned to the soil." To meet these objectives "all food waste and bio-degradable waste shall be

37. Mani Shyamala K., *et al.*, (2002) *Combating Solid Waste – A Fruitful Strategy*, pp. 2, 14, Centre for Environment Education, Delhi.

composted, recyclable waste shall be passed on to the recycling industry and only rejects shall be landfilled in a scientific manner."[38]

If the Rules did not emphatically endorse the model of household-NGO cooperation, for final disposal of over 50 per cent of municipal waste it was due to the constraint of formulating a framework applicable to 300 cities at different stages of social preparedness in the matter of pursuing a policy for processing and reducing waste which was low cost, environmentally benign labour intensive but technologically unimpressive.

The Rules state mixed waste containing recoverable resources "shall follow the route of recycling. Incineration with or without energy recovery including pelletisation can also be used for processing wastes in specific cases. Municipal authority or the operator of a facility wishing to use other state-of-the-art technologies shall approach the Central Pollution Control Board to get the standards laid down before applying for grant of authorisation."

However, the fact remained that waste could be minimised only up to a point and landfill sites were needed to be planned and prepared for in advance for waste that could not be composted or recycled and for residues from the recycling plants. In Schedule III, the Rules lay down the criteria for selection of landfill sites, the sizes in the context of waste generation for at least 20 subsequent years, minimal distance from habitation clusters, forest areas and airports. Certain on-site facilities are indicated such as a weighbridge, waste and pollution monitoring equipment and fire protection devices. There are specifications for land filling-how waste should be spread in thin layers using compactors and covered with 40-65 cm of soil or construction material. Prevention of pollution of groundwater through diversion of storm drains and liners at the base of the sites are suggested for both old and new sites. Water quality standard for the site area are laid down as also for ambient air quality. Provisions are indicated for tapping the methane gas if produced in viable quantities for power generation.

The Rules require the municipal authority to furnish an annual report as per Form I which requires exhaustive details about houses covered for collection, types of equipment used for storage, number

38. *Solid Waste Management in Class 1 Cities in India, op. cit.*, p. 7.

of vehicles used for transportation and manner of composting etc. to be submitted to the District Magistrate who in turn is required to send it to the State Pollution Control Board. The latter has to submit a consolidated report to the Central Pollution Control Board by 15 June every year.

To optimise the benefits of any endeavour to improve waste management practices, the Barman Committee Report had suggested the need to not only involve NGOs but also private sector participants. The Barman Committee recommended that the Contract Labour (Regulations and Abolition) Act, 1970, which prohibited contracting out of any service which is provided by the Urban Local Bodies (ULBs) through its own staff, be amended to facilitate this.

The Barman Committee felt that institutional strengthening was the key to success of the SWM system; Municipal Commissioners needed to have three years tenure; there needed to be more environmental and public health agencies and adequate safeguards for the supervisory staff against abuse of the Scheduled Caste/Scheduled Tribe (Prevention of Atrocities), Act of 1984 through suitable amendment. The establishment of a Management Information System was the key to monitoring of the performance of manpower and machinery.[39]

Finance was indeed a constraint in improving services but there was considerable scope for cutting down wasteful expenditure. The existing rates of taxes and charges needed to be reviewed and revised in line with the current cost of providing these services and a link up with the cost of living or inflation index for automatic annual increases. Fiscal autonomy could be enhanced through issue of tax-free municipal bonds. Unspent grants by the Tenth Finance Commission could be made available to the ULBs for modernising their SWM practices.[40]

The Barman Committee felt that there was an urgent need to Reduce, Re-use and Recycle waste. For reducing waste, incentives needed to be provided to recycling and composting industries. All types of manufacturers must be persuaded to use re-usable packaging material and incentives or product discounts given to consumers for

39. *Ibid*, pp. 63, 65, 66.

40. *Ibid*, p. 72-77.

the return of packaging or bottling materials. Public awareness was the key to success in these efforts. Public awareness campaigns using the Information, Education and Communications (IEC) techniques were required to promote the two campaigns of Waste Segregation and of Reducing, Reusing and Recycling waste.[41]

Hygienic SWM needed to be on the national agenda. Prevention of open defecation and managing cattle nuisance were also facets of SWM, the Committee felt.[42]

Taking a leaf from the Bajaj Committee Report, called for a Technology Mission under the Ministry of Urban Development for five years to monitor the performance of various local bodies in the implementation of the Committee's recommendations and to guide and assist them in adopting suitable technologies for processing and disposal of waste.[43]

Although plastic waste represents only 6 per cent of the total MSW in Delhi, it is among the biggest contributors to the solid waste problem. It is estimated that Delhi "generates 250-300 metric tonnes of plastic waste per day, half of which is recycled into carry bags."[44] To regulate the health hazards of the fast growing sector of plastic waste, the Central government notified the Recycled Plastics Manufacture and Usage Rules, 1999. The Rules prohibited usage of carry bags or containers made of recycled plastics for certain purposes. "No vendor shall use carry bags or containers made of recycled plastics for storing, carrying, dispensing or packaging of foodstuffs." Carry bags and containers made of plastics could be manufactured if they were made out of virgin plastic in natural shade or white. Carry bags and containers made of recycled plastics and used "for purposes other than storing and packaging foodstuffs" could be manufactured using pigments and colourants as prescribed in the Bureau of Indian Standard (BIS) list entitled "List of pigments and colourants for use in plastics in contact with foodstuffs, pharmaceuticals and drinking water." Recycling of waste was to be done strictly in accordance with the BIS standards and the carry bags manufactured out of recycled

---

41. *Ibid*, pp. 83-84.
42. *Ibid*, p. 79.
43. *Ibid*, p. 88.
44. Mani, Shyamala K., *et al., Less Plastics for a Better Future, op. cit.*, p. 8.

plastics were to be "marked as recycled along with the indication of the percentage of use of recycled material."

The Central government did not rest content with the enactment of Recycled Plastics Manufacture and Usage Rules, 1999. In January 2001, the Ministry of Environment and Forests, Government of India constituted the Ranganath Mishra Committee on Plastic Waste to examine various regulations on plastic waste disposal brought by the Central and State governments and to suggest appropriate measures for collection, segregation treatment and disposal of plastic wastes. Regarding recycling, the Committee noted that the Recycled Plastics Manufacture and Usage Rules, 1999 did not provide for seizure/ confiscation of illegal material nor for compounding the offence. As prosecution takes a long time, the Committee was of the view that a compounding provision in the Rules was advisable. The Committee's most significant recommendation was that the plastics industry should be made responsible to retrieve the empty packaging. Its "directive to the mineral bottle industry to collect 60 per cent of its over one billion wastewater bottles over a period of one year" has had the result of forcing the industry "to install collection centres and provide an incentive of 25 p for bottle-take-back."[45]

The largest end-user of plastics in India is the packaging industry. According to industry estimates, "plastics have captured 40 per cent of the total packaging market in India."[46] By 1998, there was already in existence, a Central governmental intervention to check proliferation of plastic packaging waste. National Plastic Waste Management Task Force was constituted by the Ministry of Environment and Forests, Government of India in August 1997 to propose incentives/penalties for checking the growth of plastic packaging waste and to prepare guidelines for packaging using plastic material. The recommendations of the task force regarding recycling were based on the BIS standards that classify post-consumer waste as Type I, post-consumer clean plastic waste of known origin, and Type II, post-consumer plastic waste of unknown origin having visible impurities, (including recycled plastic waste with visible contaminants). The task force recommended that beyond Type II materials, recycling plastic waste should be

45. Srishti, *Report on Recycling Responsibility, op. cit.*, p. 21.

46. Toxics Link Information, "Breaking the Plastic Mould," (June 2001), Information Sheet No. 13.

banned except for energy recovery. Consumer items such as toys, water bottles, carry bags etc. should not be allowed to use recycled plastic waste beyond Type I. Industry was urged to give a lead in "channelising maximum post-consumer plastic waste for recycling" by supporting schemes for collecting post-consumer plastic waste and by providing incentives to ragpickers and NGOs for collection of waste from public places. Governments were urged to promote the right type of recycling by providing technical and financial assistance and fiscal incentives to encourage manufacture of plastic products with appropriate recycled content.[47]

Delhi-21 Perspective

Municipal solid waste, according to the DUEIIP Report, Delhi-21, was the most visible component of solid waste whose other two aspects were hazardous and biomedical waste. With the all-India rules in place on management and handling of municipal solid waste, hazardous and biomedical waste, "there was no shortage of legislation governing solid waste management." The challenge before the public authorities in Delhi as elsewhere lay in enforcing the laws so that there was "phased improvements towards nationally accepted targets."

Delhi would need to dispose of between 17,000 and 25,000 tonnes of waste in 2021 even if waste reduction strategies were in place. Since land filling was the most cost effective strategy for safe disposal (Rs. 300 a tonne as compared to Rs. 3000 for incineration), the issue of identifying new sites in the NCR had priority. Confining the search to the National Capital Territory alone was unrealistic. Since land was scarce. The endeavour should be to minimise waste for final disposal through investment strategies. "Downstream changes in reduction and disposal may act as a key to spread change upstream .... the investment programme will force these synergies." Lessons from the experience of other cities such as building of compost plants at landfill sites creation of arrangements for successful marketing of compost needed to be applied in Delhi. "Private sector participation (PSP) must be used as an agent for change." PSP did not mean privatising solid waste management. It meant, "using 'third sector' organisations such as CBOs/NGOs as well as the commercial sector" and in this context

47. Srishti, *Report on Recycling Responsibility, op. cit.,* p. 23-24.

"such initiatives as exist in Chennai, Bangalore and other cities" could serve as a model for Delhi.

Delhi-21 warned against the dangers of entering into PSP agreements without adequate safeguards. Such agreements promised to bring in capital finance and management skills while reducing the municipality's direct responsibilities. "They are not, however, a free lunch." The municipality needed to develop first the capacity to "negotiate toughly and manage expertly with their contractors who they will pay." In a more sustainable scheme for solid waste management, the MCD would need to collect adequate charges for a public service while offering incentives for waste minimisation and developing and enforcing laws against fly-tipping (illegal dumping).[48]

To undertake new responsibilities and exercise its traditional functions in a more efficient and flexible manner, MCD would have to reduce or retrain its personnel through a programme of institutional and managerial reform for which there was substantial scope. Out of its 108,000 employees, 44,000 were sweepers and out of Rs. 830 million a month spent on salaries, Rs. 500 million were spent on the salaries of the sweepers. A sweeper's productivity was 0.15 tonnes a day. A generous voluntary retirement scheme was an option MCD could consider. In addition, the MCD taxation policies especially the levy on property could be exploited to augment its resources.[49]

According to Delhi-21, for sustainable management of municipal solid waste, as in the case of other services, the existing relationship between the governing authorities in Delhi needed to be modified. If control over land was transferred to GNCTD and MCD split into eight revenue districts, then the task of collection of solid waste and its transportation to the transfer stations would have to rest with the respective municipality areas. For its subsequent movement to the landfill site, the responsibility would be with the unit for Municipal Affairs in the Urban Development Department of Delhi Government.[50]

---

48. *Delhi-21*, Delhi Urban Environment and Infrastructure Improvement Project (DUEIIP) Report, (January 2001), Government of India, and Government of National Capital Territory of Delhi, pp. 49-53.

49. *Ibid*, p. 13.

50. *Ibid*, p. 100.

Interventions of Public Authorities (1998-2003)

Steps taken by the MCD

On 16 July 2002, in a meeting held at the India Habitat Centre, New Delhi on the subject of "An Interface with Civic Authorities", the two panellists, Ramesh Negi, Additional Commissioner (Sanitation), MCD and Ravi Das, Director (CSE) sounded confident about the abilities of the MCD in meeting the challenges posed by the requirement to implement the MSW Rules 2000 from 1st January 2004. These rules in essence were a prescription for sustainable solid waste management because they envisaged a durable arrangement for cooperative action between government personnel, citizen groups, private sector, and NGOs for handling and disposing waste so that there was minimal adverse impact on health and environment within costs considered affordable. Their optimism was based on the successful initiatives that had been taken to ensure better use of existing manpower, equipment, vehicles and greater public accountability. Further, contracts had been entered into with private agencies by which the MCD was legally freed from its traditional functions of maintaining dustbins, public toilets and transporting waste in some areas while being able to generate funds for discharge of its various functions. The MCD had introduced night sweeping in 40 commercial areas and it aimed to extend it to 60 commercial/ market areas by 2007. The MCD also planned to "restructure" sweeping operations where sweepers from regularised colonies were to be moved to unauthorised colonies.

The MCD was also seriously following its policy for engaging substitute *Safai Karamcharis* in lieu of *Safai Karamcharis* (*SKs*), absent with or without leave up to 10 per cent of the number of absentee *SKs* in each zone/ward/area. Hiring of casual labourers regularly in lieu of the missing *Safai Karamcharis* had considerably reduced the impact of routine absenteeism of workers on its sweeping operations. The *SKs* had been given handcarts, rickshaws and protective equipment such as gumboots and gloves. Door-to-door collection of garbage in 326 colonies/pockets had started with the help of satellite vehicles/trucks.[51]

---

51. Municipal Corporation of Delhi, *Background Note on Conservancy and Sanitation Engineering Department*, p. 3.

200 covered trucks were being purchased conforming to the Euro-II emission norms for transportation of garbage to the landfill sites and arrangements were being made with their manufacturer to ensure their long-term maintenance. There were 26 bulldozers at the three landfill sites to spread, dress, level and compact the garbage.[52]

In each zone, a control room had started functioning with a Sanitary Task Force, equipped with a wireless and consisting of a Sanitary Inspector, eight *Safai Karamcharis*, a truck and a loader, which could be despatched for immediate removal of garbage on request. On an average, according to MCD's assessment, complaints were being attended to in a period of two-three hours. Town Hall, the MCD's headquarters, was monitoring this service and responsible for ensuring that the complaints were attended to on the same day they were made to the MCD.[53] In addition, the MCD had also set up an information database as also a research and development cell.[54]

Private entrepreneurs on a Build Operate Transfer (BOT) basis had constructed/were constructing about 20 toilet blocks in commercial/ appropriate locations out of the 48 toilet blocks proposed to be privatised. Walls of the toilets were to be used as advertising space. The concession for operating these toilets was granted for five years and the facilities were to be utilised by the public on "pay and use basis."[55]

By August 2002, the Corporation also planned to privatise 315 garbage dumps out of its 2,400 dumps, most of which had outlived their utility. As per this scheme, the private operator was to maintain the dump and transport garbage to the landfill sites in return for the money that he would earn by putting up advertisements on the four walls of dump. "The private operator will pay the MCD Rs. 1,500 every month. The Corporation will save Rs. 30,000 which it spends for transportation." The lure of strategic advertising space had ensured that MCD's invitation for constructing *dhalaos*, maintaining them and transporting waste from them regularly to the landfill sites—all after substantial payment to the MCD—had attracted many takers. 25 dumps in the city had been privatised and 12 had become operational.

52. *Habitat Focus*, (2002), Newsletter of the India Habitat Centre, No. 2, August, pp. 2-3, New Delhi.
53. Municipal Corporation of Delhi, Background Note, *op. cit.*, pp. 3-4.
54. *Habitat Focus*, *op. cit.*, p. 3.
55. Municipal Corporation of Delhi, Background Note, *op. cit.*, pp. 8-9.

In addition, the MCD planned to privatise transportation of garbage to the landfill sites. To begin with the scheme would be introduced in Rohini and Shahdara zones. The cost was expected to be less than the expenditure currently incurred. "The Corporation spends Rs. 40,000 for maintaining each truck every year."[56] The MCD awarded contracts for setting up a processing plant at the landfill site at Ghazipur for converting slaughterhouse waste into chicken feed; at Narela-Bawana Road for converting waste into compost; and at Okhla for making fuel pellets out of waste. The MCD also began negotiations with private firms for putting in place arrangements for gasification, and bio-methanisation.[57]

In connection with the implementation of the Municipal Solid Waste (Management and Handling) Rules, 2000 MCD's view was that apart from finalising location of additional sites there were three distinct areas requiring special attention if it was to be held accountable for abiding by the MSW Rules in the near future.[58] These were re-mediation of existing landfill sites, initiation construction of engineered landfills and and making arrangements for ensuring deposition of segregated waste at the public collection centres.

Projections under each head were made to the Planning Commission in the context of the Tenth Five Year Plan. In addition, the MCD also took specific action in each of these areas. In the first case, since additional sites could not be secured it was decided that the levels of the existing sites at Bhalsawa, Ghazipur, GT Karnal Road and Okhla would be raised and their lives extended. Technological interventions were considered necessary to prevent sudden explosions because of accumulation of methane gas in the garbage piles. By 2003, in-house environment assessments were made and private parties identified for installing the necessary facilities. This was indeed a heartening development for citizens living in the residential areas near the landfill sites. In February 2002, there was a press report about Bhalsawa which stated that the stacked garbage in the land fill had become 20 feet, higher than the two storey houses in the nearby

---

56. *Habitat Focus, op. cit.*, pp. 2-3.
57. Tenth Five-Year Plan of Delhi, *www.delhiplanning.net.in*, pp. 1474,1475,1477.
58. Municipal Corporation of Delhi, Background Note, *op. cit.*, p. 2.

colony.[59] Clearly, a repeat of the Payatas tragedy in which 137 people had died on 10 July, 2000 "under a collapsing mountain of trash" at a dumpsite in Manila could happen in Delhi.[60] According to the press report, the mandatory requirement of six to twelve inches of earth after every one two metres of garbage height was not being followed. Space for incoming waste seemed to be made simply by setting fire to waste mounds which were considered to consist of a mix of domestic waste slaughterhouse waste, industrial scrap and thermal power plant waste.

In connection with the finalisation of location of new sites three new sites were identified. By 2003, at Jaitpur construction of an engineered landfill was at an advanced stage and the Infrastructure Development Finance Company, (IDFC) Chennai had been allotted the task of carrying out an environment impact assessment at Bhatti Mines and Narela-Bawana Road for final decision about their suitability as engineered landfill sites.[61]

Deposition of waste in segregated form as bio-degradable and non-bio-degradable categories involved an attitudinal change that the MCD knew was a slow and difficult process from its own earlier unsuccessful experiment of distributing coloured bags for collecting household waste. The Delhi Plastics Bag (Manufacture, Sale and Usage) and Non-Bio-degradable Garbage (Control) Act, 2000, which had been made effective from October 2, 2001 in the National Capital Territory of Delhi was made applicable in 114 colonies in the jurisdiction of the MCD. There was a provision for making the local authorities (MCD) responsible for providing adequate number of separate dustbins for deposit of biodegradable and non-biodegradable garbage. A fine of Rs. 5,000 or imprisonment up to one month, could be imposed for improper disposal. 217 receptacles with three coloured drums for bio-degradable, non-bio-degradable and glass/other inert material were placed in the notified colonies in the Central zone and South zone. NGOs were contacted to act as facilitators between Resident Welfare Associations (RWAs) and ragpickers to devise

59. Tankha, Rajkumari (2002). "Trouble Piles up at Bhalswa," *Hindustan Times*, February 19, Delhi Edition.

60. Roy, Barun (2000). "A Lesson for Trash Managers," *Business Standard*, July 21.

61. Khandelwal, P.K., Executive Engineer, Municipal Corporation of Delhi, Interview with the author.

income generating mechanism for the ragpickers but the efforts ended because the MCD felt that the fees that the NGOs were asking for were "too hefty."[62]

To the MCD officials, their view that "the NGOs were there only for the money" was further strengthened when as part of their preparation for the solid waste management regime it was decided to test the feasibility of a seven level slab of payment by householders in return for the service of regular collection of waste by NGOs in six of the twelve zones of the MCD. Its advertisement in the national newspapers attracted responses only from unknown NGOs from other States. Discussions with 50 to 60 of short listed NGOs broke down and the award of contracts was put on hold because the NGOs seemed interested only in getting the money without making any firm commitments about purchase of equipment and vehicles for collection of waste. Against this background, the MCD decided to explore the possibility of a partnership with private contractors who could assume full responsibility for collection of segregated waste from the householders and relieve the MCD of this duty.

Towards the end of October 2002, Rakesh Mehta, Municipal Commissioner, Delhi announced that the MCD had signed an agreement with Chennai based, Infrastructure Development Finance Company Ltd. (IDFC), to identify how precisely waste management could be streamlined in six of the 12 MCD zones. The zones included in the scheme were Central Zone, South Zone, City Zone, West Zone, Sadar Pahar Ganj Zone and Karol Bagh Zone. The inclusion of the Central and South Zones in the scheme was very surprising because in these two Zones of the MCD the NGOs had already undertaken many pilot projects and acquired considerable experience and expertise in waste segregation, collection and disposal. The MCD's thinking on how foreign companies knew best how Delhi could manage its waste in a cost effective manner is evident in the statement that Rakesh Mehta made to the press. "The IDFC shall work out the various alternatives involved in each activity of the solid waste management system and optimise it through financial planning. The IDFC will outsource activities by inviting global tenders and

62. Municipal Corporation of Delhi, Background Note, *op. cit.*, p. 9.

bidding process, design draft agreement documentation, evaluation and execution of the contracts with the parties, by experts." The entire task was expected to be completed within 45 weeks.[63] Rakesh Mehta hoped that the best national and international companies would come forward to participate in the capital's solid waste management."[64]

In April 2003, on the basis of a study done by the IDFC, Rakesh Mehta, the Municipal Commissioner, announced that in the six zones, private companies would collect the garbage from the homes, store and transport it to the landfill sites. The private operators would be responsible for the design and maintenance of a separate collection mechanism for the solid waste and its transportation to sanitary landfill sites. The Municipal Commissioner said that in the pre-transportation stage, "the operator would be designing, procuring, operating and maintaining a system including equipment and vehicles for solid waste collection from door-to-door to the *dhalaos* and transfer stations in a segregated manner." The operator would also be "redesigning the existing *dhalaos* and dustbins to allow for segregation and storage of waste, (besides undertaking the) designing and construction of transfer stations/sorting and recycling centres." For the last stage of waste management, he would be "designing a transportation system and equipment to allow for transportation of waste in covered containers with a colour coding for bio-degradable and recyclable and other dry waste." The private companies would be investing about Rs. 120 crore for this purpose over eight years, the period of the contract. Three of the waste management companies that had tendered for the awards of this work were from France, Taiwan and Singapore.[65]

While the MCD stood to gain immensely in the near future from its privatisation initiative, thousands of waste-pickers, estimated to be about 100,000 in Delhi, besides employees of recycling industries faced the prospect of a huge reduction in their income. This was

---

63. "MCD-IDFC gear up for garbage mission," *The Pioneer*, October 21, 2002, Delhi edition, *www.sdnp.delhi.nic.in/headlines*.

64. "Waste management to go into private hands," *The Hindu*, April 2, 2003, *www.sdnp.delhi.nic.in/headlines*.

65. "Waste management to go into private hands," *The Hindu*, April 2, 2003, *op. cit.* and Chhabra, Rahul "Can MCD pay for its garbage, firms wonder," *The Times of India*, April 17, 2003, Delhi Edition.

highly ironical because, as studies have estimated, the waste-pickers by handling between 9 per cent to 15 per cent of the 7,500 tonnes of waste generated daily in Delhi save "the three municipalities in Delhi, a minimum of Rs. 6 lakh daily" in labour and transportation costs.[66]

Further, MCD's initiative meant marginalisation of Delhi's best known solid waste management NGOs when the historical moment had finally arrived, as it were, to enable them to expand their activities and match and even surpass Bangalore and Chennai which were mentioned admiringly as models in the context of waste management in Delhi-21. Vatavaran, Srishti and Chintan, whose founders had 10 to 15 years experience in successfully organising grassroot initiatives in residential colonies, schools and colleges besides that of leadership in research activity and advocacy of SSWM practices, are located in Central and South Delhi. These areas were part of the zones for which the consulting firm had been hired to suggest a solid waste management solution.

However, there was soon evidence to suggest that these fears about the future were possibly not warranted. This was because the MCD signed an agreement with the Asian Centre for Organisation, Research and Development (ACORD), the NGO who was the facilitator of the Delhi Chief Minister's *Bhagidari*, the Citizen-Government Partnership, initiative to discover ways and means by which the MCD could interact with all the social groups (*bhagidars/* stakeholders) who were engaged in solid waste management operations. In consultation with ACORD, four *Bhagidari* workshops covering all registered RWAs in the 12 zones of Delhi were planned in the coming months.

The first *Bhagidari* workshop with the MCD on the Municipal Solid Waste Segregation was held from 18 to 20 December 2003 for residents of Central, South and City zones of the MCD. The three hundred odd participants included RWA representatives from each zone, teachers associated with eco-clubs in the zones, students from the ninth, tenth and eleventh standards, from the zones, MCD personnel from the levels of Sanitary Inspector, Assistant Sanitary Inspector, *Safai Karamcharis*, together with members of NGOs and professional waste workers (ragpickers) identified by the *Safai*

66. Chaturvedi, Bharati, "Waste-handlers and Recycling in Urban India," *Social Change, op. cit.*, p. 43.

Karamcharis. In the Large Group Interactive Exercise (LGIE) format of this workshop, the participants had the opportunity to dialogue and brainstorm in small-face to-face table groups of seven to ten members each and evolve common perspectives on specific problems while identifying the issues on which there were differences. It is significant that the December workshop was preceded by a preparatory workshop from 23 to 25 June 2003 involving RWAs, Delhi based NGOs, Sanitary Inspectors, Assistant Sanitary Inspectors, Sanitary Guides, Ragpickers and Union leaders of sanitary workers in which the participants became sensitised to the perceptions, concerns, views and suggestions for improvement of existing solid waste management arrangements.[67]

Initiatives of the Delhi Government

In the context of improvements of the management of municipal solid waste between 1998 and 2003, the most significant step taken by the Delhi government was linked with was institutionalisation of *Bhagidari*, the project on citizen participation in governance initiated in 2000. Its central feature was a structural arrangement called the Large Group Interactive Event (LGIE) which enabled RWAs, Market Traders Associations (MTAs), to directly confront officials of the MCD, Delhi Police, Delhi Jal Board DJB and the Delhi Vidyut Board (DVB). Instead of trading grievances and defensive positions, representatives of these entities through dialogue were able to arrive at shared perspectives and collaborative solutions about fulfilling tasks which were legally in the domain of the governing agency but whose inadequate discharge was a matter of apathy in the government officer and cynicism in the citizen. In the first phase, between January 2000 to June 2001, four workshops were held by ACORD in which 167 RWAs and 12 MTAs participated involving 358 representatives and more than 950 officials from area/zone/colonies.[68]

Seven priority areas that were related to solid waste management were taken up as part of this scheme. These were maintenance of community parks, supervision of sanitation services, information about sanitation staff, imposition of fines on littering, door-to-door

67. Asian Centre for Organisation Research and Development (2004) *Report on Preparatory Bhagidari Workshop for Solid Waste Management*, (23-25 June 2003)

68. Working Report, *Bhagidari*, The Citizen-Government Partnership, (Phase-I: January 2000 to June 2001), pp. 98-116.

collection of waste/garbage, solutions on community bins and generation of public awareness on sanitation. A nodal officer from the MCD was appointed to meet regularly with the RWAs and a cell in the Chief Minister's office was set up to track the progress of action taken on mutually accepted decisions.[69]

In the first phase, all aspects of the *Bhagidari* programme were managed centrally by the "office of the Chief Minister." The key officers here including the Principal Secretary to the Chief Minister and the Secretary to the Chief Minister personally took charge and interacted closely with the RWAs, MTAs, Heads of Departments of Civic agencies/Utility agencies as well as the process-design consultants of ACORD to personally plan and organise each step of the new initiative. In view of the organisers of the workshops, "without such personal involvement and attention to detail, no change-process can be initiated."[70]

In the second phase, which began in July 2001, to achieve greater efficiency and involve more people, the programme was decentralised at the revenue district level with the Deputy Commissioner (Revenue) as the district coordinator. The Deputy Commissioner was authorised to spend Rs. 10 lakh on organising two workshops in each revenue district. Rs. 90 lakh was provided for holding 18 workshops in the nine revenue districts. In addition, Rs. 40 lakh was provided for the Consultant's fees for conducting these workshops and Rs. 5 lakh for the documentation.[71] 450 RWAs had registered for the second phase of the programme. More than 1,100 citizens groups were successfully involved in the eight district level workshops by the end of the second phase in October 2002. Since the programme had so far attracted only the registered associations, it was hoped that in future, CBOs in the slum clusters, resettlement colonies and the unauthorised colonies and NGOs and Women groups would become *bhagidars* or stakeholders in the scheme.[72]

69. *Bhagidari*, (2002), Delhi Govt., Delhi, pp. 4-6.

70. Working Report, *Bhagidari*, The Citizen-Government Partnership, Phase-II: July 2001 to October 2002), p. 2.

71. Maitra, Shipra (2003). "Decentralised Governance—Delhi's Experience Towards Participatory City Management," Seminar on *Sustainable Development of Delhi*, New Delhi, pp. 3-4, June 16-17.

72. *Bhagidari*, *op. cit.*, pp. 8,15.

Another initiative of the Delhi government related to improvement in solid waste management lay in the fact that the Department of Environment (DoE) of the Delhi government paid Rs. 4,000 every year to 1,200 schools and colleges for the establishment and running of eco-clubs. A broad menu of activities ranging from planting trees to quiz competitions in the 3 R's, (Reduce, Reuse and Recycle) and vermi-composting projects were suggested by the Department.[73] The DoE sponsored specific imaginative environment conservation projects including those that related to sustainable solid waste management. An example was the multi-stakeholder initiative involving teachers and students of Gargi College, and an ex-student who was a Director of a Consulting and Services company. Residents of a colony near the College and the local eateries were enrolled for the project. A scheme was put in place where waste was collected, segregated and stored at source in five different bags held in a specially designed household waste holding stand which had spaces for wet/food waste, paper and card board waste, plastic waste, glass items and metal. The waste was carted from the households in a modified handcart with five compartments to a *dhalao* adapted to receive segregated waste. The waste dispatched from the local *dhalao* to the landfill was minimal because the bio-degradable waste, which forms the largest component of household waste, was composted in four containers of 700 litres capacity kept at the dumpsite.[74]

Another major initiative of the Delhi government was the passage of the Delhi Plastics Bag (Manufacture, Sale and Usage) and Non-Bio-degradable Garbage (Control) Act, 2000, on 29 November 2000, and made effective from 2 October 2001. This Act prohibited the manufacture, sale and usage of recycled plastic bags and virgin plastic bags of less than 20 microns (micron, a measure of thickness, is 1/1000 of a millimetre) and forbade the use of recycled bags for food. The Act further required that the recycled bags should be coloured and pigmented with permitted dyes and be marked as 'recycled bag unsafe for carrying food'. The name and address of the manufacturer was also required to be indicated. Violation of these provisions of the Act could involve imprisonment for a period of three months to one year and a fine upto Rs. 25,000.[75]

---

73. *4 Years* (2002). Working Report, Delhi Government, p. 47.
74. Sattar Saini (2002). "The Battle Against Waste", *Hindustan Times*, October 17, Delhi Edition.
75. *The Delhi Plastic (Manufacture, Sales and Usage) and Non-Bio-degradable Garbage (Control) Act, 2000*, Department of Environment, Government of N.C.T. of Delhi.

Chapter III of the Act forbade the throwing of any non-bio-degradable garbage or any bio-degradable garbage in non-bio-degradable bag which was likely to impair the proper functioning of drainage and sewage system and required its deposition in receptacles designated by the local authority. Violation of the provisions of Chapter III was punishable with imprisonment for a term which could extend to one month or a fine which could extend to Rs. 5,000 or both. However, provisions of Chapter III were made applicable only to 74 colonies/areas of the South Zone and 40 residential colonies of the Central Zone of the MCD; seven colonies of the NDMC; and four colonies of the Delhi Cantonment Board.

The Department of Environment of the Delhi government also launched campaigns "say no to plastic bags - say yes to jute, cloth or paper bags" to sensitise the public about the harmful effects of plastic which outweighed the convenience factor especially when low grade plastic was used for carrying cooked food. The NGOs and the constituents of the nascent *Bhagidari* programme were mobilised to launch "anti-littering and anti-plastic bag" campaigns.

The justification for these initiatives was to safeguard public health since some toxic chemical components in recycled plastic are water and oil soluble and to dry up the demand for thin plastic bags. Use of thicker reusable bags envisaged by the Act and could reduce litter even as a supply channel for the plastic recycling industry was maintained. Of course, as Dilip Biswas, Chairman, CPCB said in a press statement, a 70-80 micron thickness (the type in which new shoes are packed) requirement for polybags, which was the original idea proposed by the National Plastics Task Force in 1997 would have led to even lesser litter, because such polybags are usable many times over but the increase in the cost of their manufacture would have been so much that though it would have restricted their use it would have also "killed the industry in one go" resulting in large-scale unemployment or retrenchment of personnel who were currently engaged in the manufacture of polybags.[76] Dilip Biswas went on to add that a "more effective notification would have been a ban on smaller polybags—8.5 inches by 12.5 inches—which account for bulk

76. Gupta, Rahul (2001). "More Plastic Goes into Thicker Bags," *The Times of India*, October 3, Delhi Edition.

of the non-degradable refuse" because they are generally discarded after a one-time use.[77]

One of the main objectives of the Tenth Five Year Plan in its finalised form was the implementation of Solid Waste Management Handling Rules, 2000, Plastic Bill and other orders passed by the Supreme Court of India and Delhi High Court from time to time in various public interest litigations. A sum of Rs. 600 crore as compared to Rs. 450 crore in the Ninth Five Year Plan was secured by Delhi government for strengthening and mechanisation of conservancy and sanitation services.

Assessment

As a result of the endeavours of the MCD and the Delhi Government and the Central government between 1998 and 2003 how likely was it that non-hazardous waste would be collected in a segregated fashion, stored, transported, processed and disposed in the ideal sustainable manner that had been envisaged in the MSW (Management and Handling), Rules 2000, which were in the public domain in their draft form since September 1999 and which were required to be enforced from Ist January 2004. Could the three governmental agencies have done more so that there was less litter, less improper disposal through burning or placement in drains and discharge in water bodies and fewer adverse impacts on public health and the environment?

MCD as the governmental agency authorised to manage waste was in a far better position to do so at the end of 2003. A few months earlier it had virtually given up the effort to persuade householders to put out only segregated waste—the most crucial step in the sustainable management of MSW. It had no doubt organised separate receptacles for wet, dry and inert waste in 217 sites to implement, the Delhi Plastic Bag and Non-Bio-degradable Garbage Act 2000, but it seems as if it was convinced that such an exercise had no future. However, the organisation by the MCD, of a Large Interactive Group Event of 300 plus persons in December 2003 in which its officials engaged with RWAs, housewives, eco-clubs, ragpickers and *Safai karamcharis* in "face to face'" discussions to specifically evolve a

77. Chatterjee, Sudeshna (2001). "Bags: Micro Measure, Major Confusion", *The Times of India*, October 3, Delhi Edition.

strategy of deposit of segregated waste was a historic development. This was because the long standing view about the MCD has been that being a prototype of a bureaucratic organisation it was incapable of examining a problem from the grassroot level. As Iqbal Malik, one of Delhi's best known SWM activists and one of the earliest entrants in the field, said in 2001, "They look at garbage management as their domain and are not keen to let others near it."[78] The end result of the Large Interactive Group Event exercise was that the MCD acquired a clear perspective on how residents could be persuaded to separate all waste into 'dry' and 'wet' categories and finalise plans for ensuring publicity of the concept and implementing it in terms of modifying waste collection centres in atleast one ward in each of the 12 MCD zones by April, 2004.

An examination of the questionnaires and response sheets generated at the preparatory workshops where various stakeholders were asked about the difficulties they faced in fulfilling their responsibilities towards SWM, especially in relation to the MCD, showed that the MCD had become interested in making a paradigm shift from being the all knowing provider to a pro-active facilitator, a key requirement of the SSWM perspective. The fact that by the end of 2003 a schedule for health camps for the lower level staff had been drawn up was a good augury.

From the point of the citizen, this demonstration of a radical change in approach was a significant development because apart from comments reported in the media according to ACORD, in the context of implementing decisions arrived at through the *Bhagidari* initiative, the MCD officials were regarded as less willing to cooperate as compared with those of the DJB and the DVB. As President of one RWA said, "the nodal officers of each of the agencies concerned are to hold regular meetings but in our case the MCD simply refuses to cooperate possibly because they see it as a government scheme which must be made to fail."[79] "The overall view," of the nine Deputy Commissioners of Delhi in 2001-2002 was that, "DJB, DISCOMs and Delhi Police was quite cooperative in the problem solving process.

---

78. Malik, Iqbal, quoted in Ahmed, Kazimuddin and Jamwal, Nidhi (2000). "A Heap of a Problem," *Down to Earth*, January 31, 2000, Delhi, p. 37.

79. "RWAs Lament Civic Non-cooperation," (2001), *The Times of India*, December 22, Delhi Edition.

However, not much cooperation is reportedly received from MCD, NDMC and DDA."[80]

However, given the gargantuan task that MCD faced in implementing the new Rules in the second largest territorial area under the jurisdiction of a single municipality—Tokyo's is the largest, MCD needed to begin building up appropriate social and physical infrastructure well in advance of 1st, January 2004. CSE operations should have been decentralised to the ward level. To combat the cultural and psychological difficulties involved in a mass switch over to segregated waste disposal by adults within a limited time frame, MCD needed the help of NGOs in the field to interact continuously with community based organisations (CBOs) in residential colonies of the rich, middle class and the poor. The scope of influence of existing numbers of NGOs could have been expanded on the lines Adrian Coad has suggested in the context of his study of Mumbai.[81] After creating/or strengthening a physical base for itself in each ward, MCD could have called for registration of the RWAs from the residential areas of the ward. It could have encouraged the formation of local NGOs whose members could have been interested residents, students and teachers associated with eco-clubs in schools and colleges located in the zone to which the ward belonged. Copies of the Barman Committee Report on Solid Waste Management in Class 1 Cities in India March 1999, in Hindi translation, if necessary, could have been liberally handed out to fast forward the 'capacity building' process. Since 'learning by doing' was the only practical method of educating large numbers about new methods of solid waste management, established NGOs could have been enrolled by MCD as resource entities.

A local RWA and NGO link up could mean that through the NGO, ragpickers of the area, domestic private employees, and even family members of the MCD employees, could be engaged in primary garbage collection (door-to-door) and categorisation of recyclables in surroundings that were hygienic and safe from police harassment. In this system security considerations of the householders would not be compromised and containers meant for stored waste would not get

---

80. Working Report, *Bhagidari, The Citizen-Government Partnership*, Phase-II, *op. cit.*, p. 44.

81. Coad, Adrian (1996). "Editorial" in *Lessons From India in Solid Waste Management, op cit.*, p. 9, Government of UK.

broken by ragpickers who are likely to be part of the urban landscape for many years. This was in essence the 'Hyderabad' model of SWM where primary collection from door-to-door was done on the basis of a tripartite agreement signed between the non-governmental enterprise, residents and the municipality.[82] Under this arrangement, ragpickers and even other part time employment seekers could have been helped to supplement incomes by handling community level composting of the segregated organic component of the waste using aerobic/anaerobic methods. Households are invariably willing to pay a small fee for such projects which have been carried out successfully by NGOs in many parts of the country—by Vatavaran and Srishti in Delhi, Exnora in Chennai, Waste Wise in Bangalore and SEWA in Ahmedabad, and which are now mandated by the MSW Rules as the most cost effective and environmentally benign way for the final disposal of the bulk of the waste generated within urban limits. Additional benefits of such a community based projects, Ravi Agarwal says, apart from a high landfill diversion rate (of over 70 per cent) is that they lead to increased awareness about source segregation, help in the prevention of dumping of obsolete technology in the country and generation sustainable more secure livelihoods.[83]

The downside of boosting compost production is that one may be stuck with a product no one wants. The cost of community produced organic manure is higher than the price at which the chemical substitute is sold. The MCD was well aware of this in August 2001 when there were reports from Bhalaswa, that the advanced aerobic technology and biological innoculum based 500 metric tonne private company project which it supported was on the verge of shut down because there were no takers for its product which at rupees three a kilo was one rupee more expensive than the market substitute. The company official, Anu Khurana said she had tried to convince MCD to use her product for its parks– there were 3000 such parks, she said—but it had been unable to oblige. She said she was losing Rs. 10 lakh per month because only 10 per cent of the plant capacity was

82. Chaturvedi, Bharati (1998). *Public Waste Private Enterprise*, p. 11, The Heinrich Boel Stiftung, Berlin.

83. Agarwal, Ravi (2003). "Creating a Community Paradigm in Solid Waste Management," in *Workshop on the Need for Institutionalisation of Solid Waste Management in India*, 26 April, HUDCO, *et al.*, Delhi, 2003.

being used, since existing stocks were not being lifted.[84] Against this background, MCD together with Delhi government could have devised a strategy of providing marketing support and also taken steps to encourage additional community composting projects by NGOs. It understood very well that the technology involved was both simple and tried and tested. The real issue was its promotion and widespread acceptance. An arrangement could also have been made between the Conservancy and Sanitation Engineering Department and the Horticulture Department for purchase of compost. The MCD could have also negotiated with the DDA, NDMC and other land owning authorities, which have public parks, to lease-out land for composting to solid waste management NGOs registered with it. A mere review of the potential savings in transportation cost of garbage from these areas would have provided the MCD with adequate incentive.

MCD by this policy innovation would have serendipitously discovered a solution for two lapses for which it frequently draws flak. The first is the burning of leaves by MCD gardners. The second is the failure to round up the stray cattle from the streets. Since fresh cow dung quickens the rate of bio-degradation of leaves it will over night become a product for which there will be buyers, now and Dairy owners will get an incentive to keep the animals within their compounds.

By adopting these measures, the MCD would have got the confidence to enter into durable partnerships with NGOs to improve sanitation arrangements and SWM in slums and resettlement colonies where soild waste is more visible than in other residential colonies. However no attempt was made by the MCD to learn from what transpired in Harikesh Nagar in 1998, where within two years, the Asian Centre for Research and Development (ACORD) had been successful in putting in place a compost system covering 120 houses and 100 shops. The Delhi Development Authority (DDA) had leased the composting site to ACORD on the recommendation of the MCD, for two years. However, well before the expiry of the lease, the compost pit and the equipment shed was destroyed by the DDA

84. Chauhan, Chetan (2001). "MCD's Garbage Treatment Project Fails to Take Off," *Hindustan Times*, Delhi Edition, August 13, Also *www.excelind.com*

without any prior notice. After that, all composting stopped in the area and the garbage was dumped in the municipal bins.[85]

By 2002 there were a number of hoardings in prominent locations in Delhi which carried the message that 'no' must be said to plastic bags. In up market stores, thick paper bags were *de rigueur* for packing purchases. Media focus on school and college level eco-club initiatives and seminars and poster competitions contributed to building awareness on how plastic bags in public places were not only an eyesore but also a cause of painful death of animals and sewage backflows and drain blockages. Step by step information on composting food waste in soft drink crates and in flower-pots with the help of earthworms became available. In addition to the older ones, new SWM related NGOs acquired visibility.

Coming to structural reform, which MCD could have considered in recent years, it could have tried to extract higher productivity from the segment of the workforce on whom Rs. 500 million out of a monthly wage bill of 830 million was being spent.[86] With scope of conventional functions of sweeping, loading and unloading of waste in which 44,000 employees were directly or indirectly involved expected to reduce with more involvement of informal sector mandated the MSW Rules, at least MCD should have tried to hire the bare minimum number.

Delhi-21's suggestion for a generous VRS was not likely to be a practical option for some years. According to the Audit Report on the Government of NCT of Delhi of 2002, though legally only 10 per cent of the sanctioned strength can be hired against absentees, the working strength of *Safai Karamcharis* was more than the sanctioned strength and the excess varied between 28 per cent and 33 per cent during 1999-2001.[87] Senior officials of the MCD attribute such a situation to irresistible pressure by politicians for whom daily wage recruitment is an opportunity for exercising patronage and consolidating vote banks.[88] This pressure is mainly exercised in the field at the level of

85. Chaturvedi, Bharati (1998). *Public Waste Private Enterprise, op. cit.*, p. 12, 86. Delhi-21, *op. cit.*, p. 13.

87. Audit Report on Government of NCT of Delhi (2002), *www.cag.nit.in/audit_reports*, pp. 143-144.

88. The Energy Research Institute's Final Report (draft) No. 1999EE41, (2002), *op.cit.*, p. 180, March.

Sanitary Inspectors and Assistant Sanitary Inspectors. Besides the pressure of the local councillor, the supervisors at the field level, as Adrian Coad says, are also under great pressure from the staff they supervise. His view was that the senior officials by their periodic visits to the field could assist the junior officials to develop skills to manage the workforce they supervised. Coad's suggestion was that greater cooperation from work force leaders could be expected if the management addressed their 'felt' need. There requirements might have been very modest and involved for instance a differently designed broom or an on-site facility for repairing tools and carts.[89]

In connection with political pressure on MCD officials in Delhi, Marina Pinto, who has made an in-depth study of city governments in India says, that the interface between the Councillors and the Administrators varies at the three levels of ward, zone and the Corporation. At the ward, the politician is a "pivotal person, between the public and the municipality in the routine administrative matters." The control of the constituency fund serves to enhance his clout. "He is the man of the people, who brings the government to the citizen. As against this, the officials who are of a low level are younger, less educated, less experienced and subordinate to administrative superiors and Councillors and therefore sometimes feel threatened by them." At the level of the zone, where the interaction between the official and the Councillor is structured through a Zonal Committee and the relationship between the two is of equals, the politician's hold lessens. At the Corporation level, "the citizen-councillor and citizen-administration interface recedes into the background and it is basically the administration, represented by the Municipal Commissioner that comes on centre stage."[90] Marina Pinto's assessment is vindicated by an incident in which when a Councillor from Mongolpuri misbehaved with the MCD's Director of Sanitation, due to the latter's refusal to cancel a transfer order as per the Councillors' wishes, the Trade Unions backed the Director and 45,000 MCD employees went on a strike and said that the strike would be withdrawn only after the

89. Coad, Adrian, "Introduction," in Sneha Palnitkar (ed.) *Manual on Solid Waste Management*, pp. 4,5, All India Institute of Local Self Government, Mumbai.

90. Pinto, Marina R. (2002). *Metropolitan City Governance in India*, pp. 154-155, Sage, New Delhi.

Councillors' arrest.[91] Clearly, "political interference" as a major obstacle to better waste management arrangement could have been skillfully blocked in many cases.

The NEERI Report in 1996 made many practical suggestions to enhance work performance for institutional reform. These seem to have been by and large ignored between 1998-2003. Managers of Joint Director level should carry out work audits at various levels and an appropriate management information system for evaluation of performance should be developed. The 2002 Audit report on the Government of NCT of Delhi noted that the MCD still prepared its daily report on removal of garbage by obtaining messages through telephone or wireless. Apart from this, there was no mechanism for crosschecking or ensuring cleaning or removal of garbage. A test check of the reports of the offices as also the information received from the Residents/Traders and Welfare Associations suggested that the veracity of the reports received was doubtful. There was the same gap between reality and claim made in its records for the time taken for attending to complaints regarding sanitation services. The complaints received were recorded in the registers and once these complaints were communicated to the concerned Sanitary Inspectors, these were noted as being 'done' without checking with the Sanitary Inspectors whether they had been attended to. Instead of the 24 hours claimed by the MCD, the actual time taken to attend to complaints was 2-15 days.[92] Thus from the point of view of greater accountability and efficiency, there was considerable scope for fresh initiatives by senior MCD officials for which no new training courses in "capacity building" were required.

From the text of the Tenth Five Year Plan of Delhi, it seems that the MCD was convinced that "Delhi city has no alternative except to opt for the latest processing technology for disposal of the waste." In its view inadequate land was the biggest constraint that the MCD faced in it's functioning. Since the setting up waste processing plants was capital intensive, the MCD was considering giving financial incentives/subsidies to private entrepreneurs for setting up such

---

91. Ravi Das, Director, Conservancy and Sanitation Services, MCD, Interview with the author.

92. Audit Report on Government of NCT of Delhi, 2002, *www.cag.nit.in/audit_reports*, pp. 144-145.

processing plants.[93] The MCD could not be faulted on legal and ecological grounds for neglecting the composting option to reduce waste and for opting for tested technologically advanced alternatives through contractual arrangements with private companies such as Valorga of France and NEDO of Japan for it was permitted in the MSW Rules.

The MSW Rules explicitly provide for the manufacture of RDF (Refuse Derived Fuel) which can be in the form of pellets, liquid or gas and which is derived from combustible component of waste, from which inert material, metal and glass has been separated. A matter of concern however was on what might happen if the moisture content in the segregated MSW was higher than what the private companies had expected and the fuel extraction was not as profitable as had been thought of originally. The unique predominantly organic nature of a low income country's urban waste was emphasised in a World Bank study in 2000. In the developing countries such as India, while the organic component of waste was expected to grow from 40 per cent to 60 per cent, plastic was expected to grow only from 4 per cent to 6 per cent while paper was expected to grow substantially from 5 per cent to 15 per cent.[94] In Delhi, as in other Indian cities, given the established institution of the *kabari* or the itinerant paper and plastic waste buyer whom householders, offices and commercial institution's managers periodically patronised, any waste that was made available to the municipality was likely to have much lower paper and plastic content than in the Western cities even as paper and plastic waste generation levels became higher.

No lesson was apparently drawn from a recent experiment on waste reduction which related on incineration. In the 1980s, a Danish waste incinerator plant set up at a cost of Rs. 44 crore in Timarpur in Delhi ran for only one week because "the calorific value of the fuel was less than half of what was needed by the incinerator."[95]

It can be argued MCD's initiative in 2002 to privatise the collection, transportation and disposal of 3312 tonnes of garbage or almost half of what the city generated daily was in line with

93. *Tenth Five Year Plan of Delhi*, *delhiplanning.net.in*, pp. 1474,1477.

94. World Bank 2000, quoted in Srishti, *Report on Recycling Responsibility*, *op. cit.*, p. 15.

95. Agarwal, Ravi and Sanjay K. Gupta, *op. cit.*, p. 57.

contemporary progressive thinking. A scholar of waste management systems of developing countries since the early 1980s Sandra Cointreau-Levine wrote in 1994, that for the solid waste system the "ideal arrangement may be mix of public and private service." Public solid waste systems in developing countries were "plagued by excessive staff, obsolescent equipment, cumbersome procurement procedures, inflexible work schedules, limitations on management changes, inadequate supervision and strong workers unions," attributes which fitted the situation in Delhi. In her view, when such public service agencies were placed in competition with private contractors and allowed to make changes to become competitive, the public agencies were able to match the efficiency and costs of the private operators. MCD thus stood to gain in the future. However, as she warned, "private sector participation is a possible opportunity not a panacea...in the absence of well-defined contract performance measures, enforceable contract sanctions, vigilant contract monitoring, and cost accountability" there was no certainty that "private operators would deliver at a lower cost than that of public service."[96]

MCD's handling of the intransigent contractors in 2002 hired to desilt drains did not inspire confidence. When the MCD's vigilance department discovered that payments were being made to the contractors on the basis of number of mounds of silt rather than the actual weight of silt that was carried away, the MCD stopped payments to such contractors. The MCD also decided to change the system from payment on the basis of visual measurement to making payments on the basis of weight of sewage desilted from the drains. When fresh tenders incorporating the new conditions which included reduced payments for the apparently completed jobs, were issued, ("after changing the mode of payment we spend only Rs. 7 lakh against the Rs. 10 lakh we were paying earlier even as the amount of sewage desilted doubled,"), the majority of contractors who used to do this work lost interest in taking on the desilting work and stayed away. MCD's response was to consider blacklisting such contractors and hire daily wagers to finish the work.[97] The delay in de-silting of the

---

96. Cointreau-Levine, Sandra (1994). *Private Sector Participation in Municipal Solid Waste Services in Developing Countries*, Volume 1, The Formal Sector, Abstract and p. 23, The World Bank, Washington.

97. Rana, Pradeep (2003). "Sewage Crisis Looms as Drains Remain Clogged," *The Times of India*, April 25, Delhi Edition.

sewers resulted in water logging and sewer backflows in many areas in the summer of 2002.

For a private sector company, profit which is the paramount consideration, is affected by many contingent factors. Any shortfall is likely to made up at the expense of the service for example by reducing the number of employees, vehicles, and storage facilities *etc.* This scenario is suggested by what occurred in Ibadan in Nigeria, where when the significant devaluation of the local currency between 1987-1992 adversely affected the licensed companies engaged to collect and dispose waste from select area. The companies simply stopped buying the spare parts needed for the upkeep of their vehicles and just made fewer trips.[98] Obviously, in their view, unsightly piles of accumulated garbage were a minor matter. Against this background, MCD should have investigated more fully the ways and means of overcoming obstacles of using the RWAs and NGOs to reduce waste in order to even further extend the life of existing and proposed landfills. The Working Report of *Bhagidari* on the first phase January 2000 to June 2001 makes it clear that RWAs were willing to supervise and coordinate door-to-door collection of waste and compost making and even partially fund these initiatives, provided MCD came forward with appropriate technical and financial assistance.[99] The problem MCD faced was similar to that of municipalities in other low and middle-income countries. Especially since the late-1990s, many books are available for guidance of municipal managers that deal exhaustively with administrative, attitudinal, technical, contractual and financial concerns and discuss the steps necessary to build successful coordination between government and different kinds of small and micro enterprises. These even contain sample contracts.[100]

By institutionalising *Bhagidari*, its citizen-government partnership initiative and creating a platform where heads of Departments/Civic agencies and Utility agencies and citizen groups could evolve a mutually acceptable solutions, Delhi government set the pace for a paradigm shift on the part of a city-wide authority from being a

98. Cointreau-Levine, Sandra, *op. cit.*, p. 33.

99. Working Report, *Bhagidari*, The Citizen-Government Partnership, (Phase–I: January 2000 to June 2001), pp. 98-116.

100. Hans Christiaan Haan, Adrian Coad and Inge Lardin (1998). International Training Centre of the ILO, Viale Maestri del Lavoro, Turin, Italy.

regulator to a facilitator. This was an indispensable requirement in a city where 6,000 tonnes of waste generated by over 13 million producers has to be managed every day. Delhi Government consolidated this initiative by enacting the Right to Information Act in 2001 by which by payment of fee ranging from Rs. 50 to 500 on background information on even commercial decisions could be secured. For pin-pointing responsibility in waste related decisions, this was a significant development given the trend towards Private Sector Partnership (PSP) and Public Private Partnership (PPP) as management innovations. The Delhi Right to Information Act has been failed for having two progressive features which were missing in similar Acts—these were penalty for officials who gave wrong information or delayed it inordinately. A citizen also had the right to appeal to the Public Grievance Commission if he was not satisfied with the information provided.

Coming to plastic bags, though the DoE was able to generate awareness about dangers of improper use, it was not able to make a dent on the problem of volume increase because the most commonly used small sized bag was good for only one time use. The ban on littering on the statute books since 2000 involving a 50 rupee fine was not a deterrent to improper disposal. The Plastic Bag Act meant to regulate manufacture met a similar fate because even the authorities seem to have realised how difficult it was to establish the violation of a technical conditionality. During 2002-2003, 70 samples were collected, 19 were found to be less than 20 microns and there were no prosecutions.[101] Even if use of plastic bags could not be checked, the issue of their recycling in better controlled conditions could have been taken up with more vigour. In 2000, a *Bhagidari* partnership was established with the IIT Delhi Alumni association and waste management was one of the themes identified for investigation and initiative by a specialist Core Group constituted from members of the faculty and the alumni. This Core Group could have been also asked to find a solution to this problem.[102]

If there was an advanced MSW paradigm in place in Delhi as in other cities considerable credit goes to the Central government which

101. Dhawan, Himanshi (2003). "Will Plastic Bags Really Depart," July 2, *Times of India*, Delhi Edition.

102. Working Report, Bhagidari (Phase-I: January 2000–June 2001) *op. cit.* p. 57.

extended full cooperation to the Supreme Court in giving legal status to a comprehensive policy under the authority vested into it by the 1986 Act. However, some of its actions were creating an anomalous situation. The Ministry of Agriculture's withdrawal of a subsidy on processing of slaughterhouse waste needed to be urgently restored. Incentives such as given by the Ministry of Non-Conventional Energy Sources for generation of power from non-conventional sources needed to be extended to the recycling and composting plants. Incineration and refuse derived fuel technology were a controversial option in Indian conditions. Considering that by the end of 2001, as many as 21 composting plants were under establishment by the private sector in collaboration with municipal bodies all over the country, the Central government should have come up with some scheme to support these initiatives. The Central government also needed to similarly support recycling industries in recognition of savings in natural resource that recycling industries ensure when they use feed stock derived from waste. In the US, buy-centres are given a payment for every tonne of waste that they can demonstrate is recycled back to industry and hence saved from land disposal. Composting plants and recycling industries in India also deserved some form of financial incentives such as low cost loans, loan guarantees and tax exemptions.[103]

Thus at all levels of government, local, city and central, there was much scope for additional measures for a more sustainable management of non-hazardous waste during 1998-2003.

---

103. "New strategy to tackle garbage disposal," (2003), *The Hindu*, June 22, New Delhi.

4

Hazardous and Biomedical Waste

HAZARDOUS WASTE

Nature of Hazardous Waste

Hazardous waste is waste that is toxic, corrosive, poisonous, inflammable and explosive due to the presence of chemicals in it. Hazardous Wastes (Management and Handling) Amendment Draft Rules, 2002, define hazardous waste to mean, "any waste which by reason of any of its physical, chemical, reactive, toxic, flammable, explosive or corrosive characteristics causes danger or is likely to cause danger to health or environment, whether alone or when in contact with other wastes or substances."[1] Hazardous waste therefore requires special care in handling, storing, transporting and disposing so that human beings and the natural environment are isolated from contact with it. In cities of low income countries though relatively very small amounts of it are co-disposed with municipal solid waste, their neglect can lead to huge adverse impact on land water bodies and air which in turn can be very damaging to human health. No durable improvements in existing conditions for its satisfactory and safe disposal of municipal solid waste can come about without considering issues connected with progress from the time of origin till the point of final internment for hazardous waste as well.

Problem of Hazardous Waste in Delhi

In Delhi, industrial production over the last two decades (1980-1999) has grown four times in size and so has proportionately the quantum of waste. A feature applicable to most industries has been the manner of expansion—it is through unregulated growth with

---

1. Centre for Environmental Law Education, Research and Advocacy, *ceeraindia.org/documents*.

many more coming up in the non-conforming areas rather than in the designated 28 authorised industrial estates. Even after the major shifting of environmentally hazardous production activities and polluting units from non-conforming zones between 1996 and 2000, more than 71,000 units, were still in the non-conforming areas as per a Municipal Corporation of Delhi survey of December 2000. These accounted for more than 85 per cent of the total industrial units in Delhi.[2]

Though the nature of industry is diverse, manufacture of food, metal, textile, paper and paper products, rubber has had prominence and so have repair services. These industries are engaged in chemical processes like pickling, electroplating, anodising, textile washing and dying which are pollution-intensive and involve significant waste producing processes.

There are different estimates about the quantum of hazardous waste that is generated in Delhi. The National Productivity Council estimated it as 60,000 tonnes per year in 1997. In 2000, Ministry of Environment and Forests, Government of India put out the figure of hazardous waste as 59,423 tonnes per year, which is about three per cent of the MSW that is produced in the same period. This variation is due to the uncertainly in the number of manufacturing units and also conflicting assessment about the character of waste these generated.[3] According to the High Power Committee (HPC), set up by the Supreme Court on the hazardous waste situation in India, the figures of hazardous waste generated are based on the reports provided by State Pollution Control Boards, who in turn based their assessment on figures furnished by the industry rather than on their own inspection and verification.[4]

According to the Census of Industrial Units in Approved Industrial Areas of Delhi, carried out by the National Council of Applied Economic Research (NCAER) in 2000, there were about 21,627 industrial units in the 28 approved industrial areas in Delhi. Of these

2. State of Environment Report for Delhi 2001, (2001), prepared for the Department of Environment, and Government of National Capital Territory of Delhi, by Tata Energy Research Institute, p. 26.

3. *Ibid*, pp. 67, 97.

4. *High Power Committee Report on Management of Hazardous Waste* (2001), *cpcb.delhi.nic.in/hpc report*

industrial units, about one-fifth units were either closed or, their allocated plot was lying vacant. Nearly two-thirds of all industrial units were located in the six larger industrial areas of Anand Parbat Industrial Estate (17.23 per cent), Mayapuri Industrial Area (15.10 per cent), Okhla Industrial Area (11.34 per cent), Narela Industrial Area (9.59 per cent), Wazirpur Industrial Area (7.70 per cent) and Kirti Nagar Industrial Estate (6.82 per cent). Only one-third of the units were located in the remaining 22 industrial areas.

These industrial estates house a large number of industrial units resulting in very little space left for handling, storage and treatment of water/solid waste generated from the units. Very few Effluents Treatment Plants were found in the units itself and Common Effluents Treatment Plants (CETPs) were equally rare in these industrial estates. It was found that only about 38 per cent (on an aggregate) of the industrial units in the survey clusters had installed ETPs on their premises for treating liquid waste. Among all industrial areas, the situation in the Mohan Cooperative industrial Area was particularly bad, as only 2.16 per cent of the units in this area had installed ETPs. The study also found that a large number of the installed ETPs was not conforming to the required standards and therefore not solving the purpose for which they had been installed. It was further observed that oil grease traps, which had been installed by some units, were ineffective in ridding the disposed water from oil/grease components. Similarly, although bar screens had been installed, they were found to be broken. The Census also found that the units were also not financially in a position to install expensive equipment to treat and dispose of the hazardous waste.[5] The pattern of indiscriminate disposal of hazardous waste in Delhi seems to be similar to that of other cities. HPC had found that few industrial units were enlightened enough to provide secured on-site facilities within the boundaries of their properties and most of the industries discharged their hazardous wastes in illegal dump sites outside industrial estates, along roadsides, in low-lying areas, along with municipal wastes or even in river and canal pits.[6]

---

5. *Report on the Census of Industrial Units in Approved Industrial Areas of Delhi*, (2002), National Council of Applied Economic Research, Delhi, *ncaer.org*

6. *High Power Committee Report on Management of Hazardous Waste, op.cit.*

Social and Environmental Implications

By mid-1980s, various studies were showing that water and air pollution was increasing at a rapid pace in Delhi and in all other metropolitan towns and there was growing frustration with the fact that the Government was doing little to control the situation.

The Bhopal Gas leak tragedy had taken place in 1984 and there was growing concern about hazardous industries that emitted toxic gases that were located in densely populated areas of Delhi. By referring to this instance and to the number of small and large chemical industries, both in the private and public sector such as Hindustan Insecticides which were operating in densely populated areas, often lacking in even minimal safety measures, M.C. Mehta filed a public interest litigation requesting the Supreme Court to close down the hazardous industries/units located in densely populated areas of Delhi or shift such hazardous units far away from the population. M.C Mehta's concern was vindicated within months when gas leaked from Shriram Foods and Fertilisers Limited on 5 December 1985. According to Mr. Mehta, "Thousands of people fled for safety, a large number was hospitalised and one person died."[7]

In spite of such incidents, practically nothing was done by successive governments to implement the provisions of the Master Plan of Delhi, apart from declaring certain non-conforming areas as conforming areas, till the Supreme Court in a series of judgments from 1996–2000 intervened in the matter and gave extensive orders for relocation of industries. While the hazardous industries, extensive industries in residential and non-conforming areas (H category), hot mix plants, arc/induction furnaces and 243 brick kilns were moved out of Delhi or closed down, non-extensive F category units in non-conforming areas continued to exist till 2000-2001.[8]

The hazardous waste generating units pose huge public health and environment dangers because small entrepreneurs who do not have adequate capital to adopt efficient treatment technology or devise safe disposal methods run most of these units. As per a case study cited

---

7. Mathur, Kuldeep (2003). "Battling For Clean Environment: Supreme Court, Technocrats and Populist Politics in Delhi," *Seminar on Sustainable Development of Delhi*, June 17, *op. cit.*, p. 5.

8. Central Pollution Control Board, *cpcb.delhi.nic.in/legislation*

in India: State of the Environment 2001 prepared by the Tata Energy Research Institute, in the Wazirpur Industrial Estate, Shahadra-Maujpur Industrial Estate and along the Grand Trunk Road in Delhi, "small and tiny-scale industries processing ferrous metals such as copper, brass, aluminium, as well as steel rolling mills and pickling factories were dumping their heavy metal rich effluent and acids into cess pools or drains."[9] This meant that effluents such as arsenic, chromium, cadmium, lead, mercury and zinc, apart from draining into the Yamuna River, which supplies 80 per cent of Delhi's drinking water, were permeating into the water table and contaminating the groundwater that is routinely used as a supplementary source of potable water. If this was the waste disposal situation in industrial estates, one could safely assume that in crowded residential/ commercial areas (non-conforming areas), an industrial unit could have only more lethal consequences. This is because hazardous emissions are not only solid and liquid but also gaseous.

The major source of air pollution in Delhi is from the number of vehicles that is growing rapidly. Between 1990-1991 and 2000-2001, the contribution of vehicular traffic to air ambient pollution increased from 64 per cent to 72 per cent. Although the contribution from domestic and industrial sector to pollution declined in percentage terms from 36 per cent to 28 per cent in the same period, in absolute terms the emission loads have increased. Apart from the rise in consumption of petrol and diesel, the two automobile fuels, major sources of pollution are fuel combustion in industrial activities and domestic stoves. Among the large-scale industries, the three coal based power plants operating in Delhi are the major contributors to air pollution. The contribution to air pollution by more than 126,000 small and medium scale industrial units, functioning from conforming and non-conforming areas as per the Master Plan of Delhi, was difficult to ascertain because many of these units between 1998-2003 were either in the process of shifting or closing down pursuant to the Supreme Court's order. However this sector was considered to be a "significant contributor to Delhi's pollution load" because of its "low combustion energy efficiency and use of inferior fuel." Domestic sector's contribution to pollution load was not significant because of

9. *India: State of the Environment 2001* (2002). Ministry of Environment and Forests, Government of India, envfor.nic.in

the increasing number of households using cleaner fuels such as liquefied petroleum gas (LPG) and kerosene.[10]

Increasing levels of air pollution are responsible for higher incidence of respiratory diseases, cancer, heart disease and nervous disorders in the city. According to one study, Delhi's polluted air is held responsible for 40 per cent of the emergency admissions of patients with breathing and heart complaints to hospitals. A number of other studies have established the association between high dust air pollution and acute respiratory symptoms thereby resulting in high health costs. According to an estimate, the number of premature deaths due to ambient Suspended Particulate Matter (SPM) in the air has increased from 7,491 in 1991-1992 to 9,859 in 1995.[11]

Government of India Initiatives

In India, 1970s and early 1980s saw the beginning of some legislative measures for the protection of the environment, particularly against air and water pollution. Water (Prevention and Control) of Pollution Act was passed in 1974 and Air (Control and Pollution), Act in 1981. However, the concern and need to manage the hazardous waste generated in the country in a scientific manner was felt only in the mid-1980s after the occurrence of the Bhopal gas tragedy on 2/3 December 1984. The Central government became acutely aware of environmental damage and the casualties that hazardous chemical substances and toxic wastes could cause. The Environment (Protection) Act was enacted in 1986 by the Central government enabling it to take any measure to improve the quality of environment and to control and reduce pollution.

The Hazardous Wastes (Management and Handling) Rules 1989 were notified on 21 October 1989 under the Environment Protection Act to specifically deal with the problem of disposal of hazardous waste. The Rules were amended in 2000, 2002 and 2003. The 1989 HW Rules provide for control on generation, collection, treatment, transportation, import, storage and disposal of hazardous waste. Wastes are not allowed to be imported for

---

10. *State of Environment Report for Delhi 2001, op. cit.*, p. 33.
11. *Ibid*, pp. 46,47.

dumping or disposal but are permitted for import for processing and reuse as raw material.

The Hazardous Wastes (Management and Handling) Rules, 1989, as amended in 2002 and the Draft Rules, 2002, categorise hazardous substances in 18 groups.[12] Some of the significant groups include:

- Mercury, Arsenic, Thallium and Cadmium bearing wastes.
- Hydrocarbons including solvents.
- Wastes from paints, pigments, glue, varnish and printing ink.
- Wastes from Dyes and Dye intermediate.
- Waste oil and oil emulsion.
- Sludge arising from treatment of wastewaters containing heavy metals, toxic organics, oils emulsions and spend chemical and incineration ash.
- Wastes from manufacturing of pesticides and herbicides and residues from pesticides and herbicides formulation units.
- Acid/Alkaline/Slurry.

According to these Rules, it is the responsibility of the 'occupier,' generating hazardous wastes, which are listed in the Rules, in quantities equal to or exceeding the limits given in the Schedules, to properly handle such waste. The occupier is also responsible for proper collection, transportation, treatment, storage and disposal of these wastes. Industry associations can themselves set up a treatment and hazardous treatment facility to deal with the wastes, in accordance with the Rules. Only 'authorised' persons can handle waste. The State Pollution Control Boards have been designated as the competent authorities to grant authorisation or permission for collection, reception, treatment, transport, storage and disposal of hazardous wastes. Authorisation will be given by the State Pollution Control Boards to accept waste only if there are satisfactory technical capabilities and equipment to handle hazardous waste. The State Pollution Control Boards may cancel an authorisation issued under these rules or suspend it if the authorised person has failed to comply with any of the conditions of the authorisation or with any provisions of the Act or these Rules.

---

12. Centre for Environmental Law Education, Research and Advocacy, *op.cit.*

Some of the other important provisions of the 1989 Rules as amended in 2000 are:

- According to Rule 3 (d), "authorised person" means a person or an organisation authorised by the competent authority to collect, treat, transport, store or dispose of hazardous wastes in accordance with the guidelines to be issued by the competent authority from time to time.
- Rule 4, which was amended in 2000, states that the occupier and operator of a facility "shall be responsible for proper collection, reception, treatment, storage and disposal of hazardous wastes" listed in the three Schedules. They should also take all steps to ensure that such wastes are properly handled and disposed of without any adverse effects to the environment. The Amendment Rules also stipulated that it would be their responsibility while handling waste to "contain contaminants and prevent accidents and limit their consequences on humans and the environment" and "provide workers on the site with information, training and equipment necessary to ensure their safety."
- Rule 5(1) provides for grant of authorisation for handling hazardous waste. It states that "hazardous wastes shall be collected, treated, stored and disposed of only in such facilities as may be authorised for this purpose" and "every occupier or recycler generating or recycling hazardous wastes shall make an application... to the State Pollution Control Board for the grant of authorisation for any of the above activities." Draft Rule 5(2) provides that "an occupier or a recycler not having a hazardous wastes treatment and disposal facility of his own and operating in an area under the jurisdiction assigned by the State Pollution Control Board/Committee for a Common Treatment, Storage and Disposal Facility (TSDF) shall become a member of this facility, pay charges as may be required" and take all other necessary steps to ensure proper treatment and disposal of hazardous wastes generated failing which the authorisation granted to the said occupier or recycler shall be cancelled.
- Rule 5(4) stipulates that the Member-Secretary, State Pollution Control Board "shall not issue an authorisation unless it is satisfied that the operator of a facility or an occupier, as the case

may be, possesses appropriate facilities, technical capabilities and equipment to handle hazardous wastes safely."

- Rule 7, as amended in 2000, says that the occupier or operator of a facility "shall ensure that the hazardous wastes are packaged, based on the composition in a manner suitable for handling, storage and transport and the labelling and packaging shall be easily visible and be able to withstand physical conditions and climatic factors."
- According to Rule 9, "the occupier generating hazardous waste and operator of a facility for collection, reception, treatment, transport, storage and disposal of hazardous waste shall maintain records of such operations" and send annual returns to the State Pollution Control Board.

Schedule 1 in the 1989 Rules that indicated 18 categories of waste was substituted by three schedules. Schedule 1 described the processes generating hazardous waste and the waste streams. Installations possessing any of the processes indicated along with waste streams would be covered by the Rules. A list of concentration limits in the constituents of the waste was indicated in Schedule 2 which would be used for classification of the waste stream as hazardous or non-hazardous. Schedule 3 contained a list of particularly hazardous substances identified in annexure VIII and IX in the convention on the Trans-Boundary Movements of Hazardous Waste held in 1989.

The Amendment Rules, 2000 introduced some significant changes in respect of the disposal sites. Rule 8 of the Hazardous Wastes (Management and Handling) Rules, 1989 provided that the State government or a person authorised by it shall undertake a continuing programme to identify the disposal sites and compile and publish periodically an inventory of such sites within the State for the disposal of hazardous wastes. By amending Rule 8, the Amendment Rules 2000, lightened the burden on the State and made the occupier or any operator of a facility responsible for identifying sites for establishing hazardous wastes disposal facility and the State government, operator of a facility or any association of occupiers responsible for identification of sites for common hazardous waste disposal facility in the state, after carrying out a proper environmental impact study. The amended Rules also prescribed that the disposal site should be set up as per the guidelines issued by the government and before setting of the facility its design and lay out should be got approved from the SPCB and operation and closure of the landfill would be as per the design approved.

Under Rule 11 of the HW Rules, 1989, applications for imports from importers were to be processed by the concerned SPCB and no imports of hazardous wastes were allowed unless authorised by the Ministry of Environment and Forests (MoEF) of the Central government. Inspite of the Supreme Court's ban in 1997 on import into India all hazardous wastes listed in List A, Annex. VIII of the Basel Convention on the Trans-Boundary Movements of Hazardous Waste, unauthorised imports of hazardous waste continued. The Rules were amended in 2000 partly to bring them in compliance with the Basel Convention. List A and B wastes of the Basel Convention adopted as Annexure VIII and IX to the Convention in 1998 were introduced as Schedule 3 of the HW Rules 2000. It was explicitly stated in the Rules that import and export of hazardous waste "shall only be permitted as raw material for recycling or reuse." Based on the recommendations of the HPC submitted to the Supreme Court, the MoEF and the CPCB drew up a new regulatory regime governing imports of recyclable wastes as per which traders were not permitted to import such wastes and importers were granted permission only after they are visited, inspected and registered for the purpose by the MoEF.[13]

The draft Rules 2002 have expanded the list of wastes to 36 streams which generate hazardous waste, make it mandatory for the recyclers generating or recycling hazardous waste to obtain permission from the State Pollution Control Board for collection, reception, treatment, transport, storage and disposal of hazardous wastes and made it necessary for recyclers of non-ferrous metal wastes and used lubricating oil to register with the Ministry of Environment and Forests before undertaking recycling. The Rules also aim to streamline the import and export policy of hazardous waste.[14]

Schedule 8 of the Rules lists 29 categories of hazardous waste, which are prohibited for import and export under all circumstances. The Directorate General of Foreign Trade (DGFT) is the agency under

13. High Power Committee *Report on Management of Hazardous Waste, op.cit.*

14. Centre for Environmental Law Education, Research and Advocacy, *op.cit.*

the Union Ministry, which issues license for all imports including scrap. The practical reality of trade in hazardous waste is that it is controlled by two sets of regulation that of MoEF's 1989 Rules as amended in 2003 and DGFT's export-import (EXIM) policy. Each regulation has its own definition of waste. The differences between the two regulations lead to confusion. So waste oil import was allowed under the EXIM policy while it was prohibited under Hazardous Waste Rules. The difference was resolved in October 2003, when the Supreme Court ordered that 1989 rules will supercede EXIM policy.[15]

There are many small units in India, which collect lead batteries for re-processing, without having any regard for pollution control measures. With a view to tackle the problem of disposal and recycling of used lead acid batteries, the Central government notified the Batteries (Management and Handling) Rules, 2001 on 16 May, 2001.

These Rules are applicable to every manufacturer, importer, re-conditioner, assembler, dealer, recycler, auctioneer, consumer and bulk consumer involved in manufacture, processing, sale, purchase and use of lead acid batteries or components thereof. The Rules make it their responsibility to ensure that used batteries are collected back as per a prescribed schedule. They are required to set up collection centres either individually or jointly for collection of used batteries from consumers or dealers; ensure that used batteries collected are sent only to the registered recyclers and that no damage occurs to the environment during transportation; and file a half-yearly return of their sales and buy-back to the State Pollution Control Board every year.

These entities have also been given the responsibility to create public awareness by all possible means about the hazards of lead and the responsibility of consumers to return their used batteries only to the dealers or deliver at designated collection centres; and buy recycled lead only from registered recyclers.

The recyclers are required by these Rules to apply for registration to the Ministry of Environment and Forests or an agency designated by it and ensure strict compliance of the terms and conditions of registration. They are also required to submit annual returns to the State Pollution Control Board and made subject to it for inspection.

---

15. *Down to Earth* (2004). "Just Dump it" Nov. 15, p. 30, 35.

The procedure for registration/renewal of registration of recyclers has been made quite stringent. The Ministry of Environment and Forests is required to ensure that the recyclers possess appropriate facilities, technical capabilities, and equipment to recycle used batteries and dispose of hazardous waste generated.

The State Pollution Control Boards are the prescribed authority for ensuring compliance of the provisions of these Rules. The Central Pollution Control Board is required to compile and publish the data received every year from the State Boards and review the compliance of the rules periodically to improve the collection and recycling of used lead batteries. Ministry of Environment and Forests or an agency designated by it is required to develop a system for computerised tracking of distribution and sale of batteries; collection, auction, transport and re-processing of used batteries; sale of re-processed lead by registered recyclers; and sale of lead from all domestic producers or importers.[16]

Delhi-21 Perspective

Delhi-21 held that since there was no reliable inventory of the hazardous waste generated by industry, an estimate that it was 30,000 tonnes or about 100 tonnes a day, could be the basis for planning a strategy for its sustainable management. Protection of health and the environment meant basically that all industrial units that use hazardous materials and chemicals dispose of their waste in a properly controlled manner and reduce the quantity of wastes that has to be disposed of by long term containment through such methods as reprocessing and reuse of waste. As per the Hazardous Wastes (Management and Handling) Rules, 1989, it was the responsibility of industrial units to store, transport and dispose of the hazardous waste.

However, since most of the industrial units were too small to organise the effective management of their waste, it was appropriate that Delhi government should through the DSIDC construct a number of Common Effluent Treatment Plants (CETPs). Since there was no facility for treating and disposing of sludge produced by the CETPs which might be toxic, necessary sites should also be identified for their safe disposal. DPCC which had the responsibility to monitor the

---

16. *Ibid.*

collection and disposal of waste and cancel licenses and close industrial units that did not comply with environmental regulations should prepare an inventory of the wastes produced. On its basis, a private sector contractor could be licensed by the DSIDC to set up a facility for ensuring the collection and final disposal of waste. The Delhi government should provide the land for this purpose. The DPCC should be responsible for monitoring and regulation of waste disposal facility. Industries producing hazardous waste and using the facility should be charged an appropriate rate depending on the volume and toxicity of the waste. Waste minimisation should be encouraged through rebates for on-site treatment of hazardous waste.[17]

Interventions of Public Authorities (1998-2003)

Efforts of the Delhi Government

The Supreme Court had given various directives from time to time since 1996 on reducing pollution in Delhi. These directives included the strict implementation of the Master Plan of Delhi 2001, closure of hazardous and extensive industries, relocation/shifting of 'F' category industries functioning and operating in residential areas/non-conforming areas of Delhi and cleaning of the Yamuna River. The Delhi government adopted a multi-faceted approach to comply with the various Supreme Court directives. Keeping in view the long-term perspective of reducing pollution in Delhi, it took a policy decision to encourage the promotion and establishment of only high-tech, low-energy intensive and non-polluting industries in Delhi. It decided to develop new industrial estates for relocation of industries operating in residential and non-conforming areas of Delhi and expedite allotment of alternative sites to such units in the existing industrial estates. Further, it decided to push for an amendment to the Master Plan of Delhi 2001 to provide for *in-situ* regularisation of various industrial concentration areas where more than 70 per cent residential premises had already been converted to industrial use and operation of the service sector industrial units, which catered to the day to day needs of the residents, from local commercial/residential areas.

---

17. *Delhi-21,* Delhi Urban Environment and Infrastructure Improvement Project (DUEIIP) Report, (January 2001), Government of India, Ministry of Environment and Forests and Government of National Capital Territory of Delhi, Planning Department, pp. 52,53.

Delhi government acquired about 1865 acre of land in Bawana, Holambikalan and Holambi Khurd to develop industrial estates for relocation of industries operating in residential and non-conforming areas of Delhi. The government also obtained the approval from the DDA and the Union Ministry of Urban Development for change of land use from Agricultural to Industrial in respect of this land. Lay out plans in respect of land acquired in Bawana for shifting of Industries, were also got approved by the DDA and the Delhi Urban Arts Commission. Development work in respect of all the five sectors in Bawana was commenced.

By December 2002, 27,500 applications for relocation were considered and 27,002 applications approved for allotment of alternative sites. Specific sites were allotted to 16,395 applicants in industrial areas such as Narela, Bawana Sector 1 and 5, Jhilmil Industrial Area, Patparganj and Badli. The Delhi government had also requested DDA on 1st October, 1999 for amendments in the Master Plan like *in-situ* regularisation of various industrial concentration areas where more than 70 per cent residential premises have already been converted to industrial use, making parameters governing house-hold industries broad-based and to allow service sector industry which cater to the day-to-day needs of the residents of Delhi such as *Atta Chakkies*, Dry-cleaners, Scooter Repair shops, to operate from local commercial/residential areas. The Delhi government also identified 24 such areas having industry concentration of 70 per cent and above for *in-situ* regularisation. In order to ensure that such areas were provided with requisite infrastructure and were developed in a proper manner, detailed project reports in respect of three areas were got prepared by experts with the expectation that these three areas could be replicated in other areas of industrial concentration also. The proposal for amendment to the Master Plan for *in-situ* regularisation of industrial concentration areas, though approved by the DDA, was not viewed favourably by the Central government. The Central government, however, did agree to the proposal to broad-base parameters governing household industries and to do away with the distinction between light and extensive industrial areas.[18]

---

18. *4 Years*, Working Report (December 2002), Delhi Government, pp. 38-40 and *Tenth Five Year Plan of Delhi, delhiplanning.nic.in*, p. 230.

In July 2004, the newly elected Delhi Assembly passed a unanimous resolution to recommend to the Central Government to declare all areas with more than 70 per cent industrial units as industrial areas to "avoid panic among the people."[19] Subsequently the Union Urban Development Ministry announced its decision to amend the Master Plan of Delhi to meet this requirement.

Steps Taken by the Delhi State Industrial Development Corporation (DSIDC)

The steps taken by DSIDC, an autonomous body under Delhi government, to reduce the health and environmental impact of hazardous waste generation had their origin in the directives of the Supreme Court. On 9 April 1996, the Supreme Court with a view to abate pollution in the Yamuna River, had ordered setting up of CETPs for the 28 industrial estates in Delhi. Half of the cost was to be borne by the government— 25 per cent by the Central government and 25 per cent by the Delhi government. The other half was to be borne by the industry, whose units could obtain up to 30 per cent loan from the Industrial Development Bank of India (IDBI).

The National Environmental Engineering Research Institute (NEERI), Nagpur, was appointed to provide technical assistance in the construction of the CETPs. Based on a survey and strategy suggested by the NEERI, the Court further ordered on 5 December 1997 that the DSIDC would build 15 CETPs for the 28 industrial estates. The cost of construction of the CETPs was estimated at Rs. 254 crore. The progress made in the construction of the CETPs was slow. The matter of construction of the CETPs assumed urgency with the Supreme Court's order of September 1999 that no industrial unit would be allowed to discharge any effluent without conforming to the standards laid down by the CPCB. The construction of the second CETP began in January 2000 and by the end of 2003, construction of only three CETPs had been completed and ten were in various stages of completion.

Societies had been registered for management of the CETPs in each industrial estate. In November 2000, the Delhi Assembly passed the Delhi Common Effluent Treatment Plants Act, 2000 to entrust

---

19. "Change Master Plan, Save Delhi" (27 July, 2004) *Hindustan Times*, Delhi Edition.

these societies with the running of the CETPs. Subsequently Delhi CETP Rules, 2001 were notified to effect recovery of the capital, operational and maintenance cost from the defaulting industrial units. The Act also called for the creation of another appropriate authority to issue notices and collect payments from individual units.[20]

According to the law, DPCC is required to regularly monitor the compliance of the Hazardous Wastes (Management and Handling) Rules by the industrial units. In 2000, closure orders were passed on 338 defaulting units in Mangolpuri Industrial Estate.

With a view to help small-scale entrepreneurs for establishing ETPs, three consultants were empanelled by the DPCC and made responsible for issuing adequacy certificates to the ETPs installed by the industrial units.

A seminar on Hazardous Waste Management was organised in January 2001 by the DPCC to create awareness about the subject and learn about recent developments in the field. Experts from Gujarat, National Productivity Council, Centre for Science and Environment, Central Pollution Control Board and all industrial units participated in the seminar.[21]

Assessment

Against the background of the SSWM perspective, as a result of the interventions of the Central government, Delhi government and the DSIDC, between 1998-2003, how likely was it that hazardous waste generated in Delhi would be more safely stored and disposed, even as there was greater protection of livelihood concerns linked with industrial activity? Also, as a result of these interventions was entry of unauthorised hazardous waste into the city likely to be reduced? Finally, given the known huge potential of damage to public health and the environment, to which public spirited individuals like M.C. Mehta have drawn attention consistently since 1985, and the media periodically—at least on the anniversary of the Bhopal Gas Tragedy—could the public authorities have done more to minimise adverse impact of hazardous waste?

---

20. *4 Years*, Working Report, op.cit., p. 40.

21. Delhi Pollution Control Committee, *dpcc.delhigovt.nic.in*

There were undoubtedly more nuanced controls in place to fix responsibility for proper disposal. The dumping of hazardous waste by developed countries even after the Supreme Court banned import of items of List A of the Annex VII of the Basel Convention into India, led to the introduction of Schedule III within the HW Rules, 2000 by which these listed substances were incorporated in the Rules itself. Also hazardous waste was to be permitted only for select recycling activities. In the matter of identification of sources of hazardous waste, the parameters were broadened to encompass not only more specific chemicals but also installations, processes and concentrations and the obligation to provide information about these was placed on the generator. By 2003 lists of hazardous substances banned for both import and export had been expanded. Acknowledging the dangers of exposure to lead, which could cause behaviour disorders at low concentration levels and mental disorder and death at higher level, the Battery (Management and Handling) Rules, 2001 were made applicable to all concerned persons who came into contact with batteries from the time of their manufacture till their final disposal.

However, there was not much evidence on the ground to suggest that the industries in Delhi were managing their hazardous waste in a more responsible fashion by the end of 1998-2003 than in earlier times. Most of these were still in non-conforming zones where waste disposal was regarded as a non-issue. Many used non-hazardous waste such as iron scrap and poly vinyl chlorides (PVC) scrap which when burnt produced tonic substances. Between 1997-1998 to 2002-2003, the quantum of iron and steel scrap import grew from 1.64 million tonnes to 2.39 million tonnes, a compounded growth rate of 45 per cent. In the case of PVC, between 1997-2003 there was a 700 per cent increase in the imports.[22] Despite the Batteries (Management and handling) Rules being in place, MoEF survey team in 2002 found clusters of illegal lead traders in Gokhale Market near Kashmiri gate, Motia Khan in Old Delhi and Gali Lallu Mir in Sadar Bazar. These areas represented the most significant link in the chain by which second hand batteries are produced. 70 per cent of the recycling of lead—Delhi is a major centre—occurs in the informal sector in the country without any pollution control.[23] It was not a surprise that in

22. "Just Dump it" *op.cit.* p. 31.

23. *Ibid* p. 34.

2001 the High Power Committee appointed by the Supreme Court to suggest improvements in the management of hazardous waste, noted that there was unauthorised in-flow of hazardous waste to recycling centres within the country. In its view the Central government had failed to put in place adequate infrastructure at the entry points of imports of such waste. The Ministry of Environment and Forests through its communication with the Ministries in charge of Ports and Customs had been unable to ensure that the ground level officers had the knowledge, authority and the technological resources to take preventative action against illegal imports.

The HPC also noted that the State Pollution Control Boards (SPCBs) had not intervened to ensure closure when industries within its jurisdiction were processing imported hazardous waste obtained without authorisation. The state government's failure to ensure adequate staff and equipment to the SPCBs contributed to an attitude of laxity.[24] In the case of the Delhi State Pollution Control Board, in 2003 in Group A, out of the five positions of Senior Environment Engineers, only three positions had been filled up. All the sanctioned posts in respect of Environment Engineers (9), Scientist-C (3), and Law Officer (1) were vacant.[25]

The HPC was of the view that the principal responsibility of the MoEF as 'the nodal agency' was to ensure that necessary concern and sense of urgency was recognised by government bodies and that there was coordination among the various central ministries and state governments. The MoEF, HPC said, had to create awareness in society and other stakeholders at large. A broad scope campaign to educate people about hazards posed by chemicals on the lines of anti-tobacco campaign was required.[26]

The HPC recognised that many industries dealing with hazardous waste, particularly lead waste and oil were in the small-scale sector which meant that they would be unable to afford costly pollution control equipment. Illegal dumping of waste was likely to be the option most commonly resorted to. In its view, CSIR laboratories had to develop and promote technology for the processing of waste

24. *High Power Committee Report on Management of Hazardous Waste, op.cit.*

25. *Delhi Pollution Control Committee, dpcc.delhigovt.nic.in,* Delhi 21, *op.cit.*

26. *High Power Committee Report on Management of Hazardous Waste, op.cit.*

through low cost cooperative solutions.[27] HPC's views are corroborated by independent analysts who however believe that the present proliferation of hazardous waste is as much due to the indifference of industry as of governmental agencies. "Industry's investment in research and development is as miniscule as government's intent to achieve pollution neutral growth: on an average it is a pathetic .1 per cent of total turnover. In addition, most Indian industrialists prefer informal solutions to structural problems, solutions such as the readily corruptible local politician or toxic sludge surreptitiously disgorged into a stream at night."[28]

Easy disposal of hazardous waste into public spaces and drains is also due to low administered price of water. "Companies pay a ridiculous ten paise for a 1,000 litres of water; in contrast people in some cities end up paying between Rs. 10 and 50 for the same quantity."[29] Water consumption efficiency of Indian companies is much below global standards. There is not much motivation for entrepreneurs in India to make, as the HPC said, production system 'cleaner' so that hazardous waste generation was minimised, or make it "clean" so that production design was modified and waste eliminated altogether. The HPC was of the view that financial incentives had to be offered to encourage these processes and the 'best practices' which embodied the precautionary principle promoted. Further, there was urgent need to examine the link between environment and health. "New types of exposure and epidemiological studies need to be carried out, including biological monitoring. It is imperative to learn about hormone disrupting chemicals and their impacts on human life and wildlife and develop the analytical facilities to analyse them in extremely minute quantities and monitor their presence in all relevant areas."[30]

The Delhi government could not be faulted for shifting industries only selectively from non-conforming zones as per the Master Plan of 2001 and excluding only the more polluting units from their existing locations, considering majority of the 1.37 lakh industries in Delhi

27. *Ibid.*

28. Bhushan, Chandra (2003). "By Hook, Crook," *Down to Earth*, June 15, p. 26.

29. *Ibid.*, p. 28.

30. *High Power Committee Report on Management of Hazardous Waste, op.cit.*

were in these spaces. According to Delhi-21, they were a major component of Delhi's economy, with an investment of Rs. 2,524 crore and an annual production worth Rs. 6,310 crore. They provided employment to a large workforce of 11.36 lakh with an average of nine workers per unit. "They use municipal and other services, pay commercial rates and keep Delhi City, as an engine of growth running to its capacity."[31] However, in the immediate future, due to the uncertainty about the timing and precise category of industries to be shifted, improper disposal of hazardous waste was unavoidable. The Managing Director of DSIDC was of the view that 16,000 plots would become ready in Bawana and Narela by April 2002 and the first industrial units would start operating there by the end of December 2002.[32] Since the Supreme Court's orders were well known, the plots could have been endowed with appropriate infrastructure well before these dates. If the Congress government had a perspective on hazardous waste that was different from the earlier BJP government, from the kind it had on public participation, as evident in the *bhagidari* initiative, it would have provided for the needs of the factory owners and workers. They in turn may have welcomed the idea of shifting there instead of mobilising all their resources to extend tenure on encroached land. There were reports from those who moved out even in 2003 that there were no proper roads and drains in the new industrial areas.

Also, in retrospect the relocation exercise could have been treated as an opportunity to completely understand the industrial processes in both the formal and informal sector. Delhi government could then have devised practical schemes for assisting industries in ensuring both better protection for workers and reduction of pollution by not only creation of common pollution control facilities but also forging of intra industry links where waste of one became raw material of another. Such a scenario is suggested by Delhi-21 where a proposal for the sustainable development of industry for the next 20 years has been elaborated. According to it, as an aspect of the locational policy, there should have been compulsory registration of all industrial units with the Industries Department of GNCTD where specially trained

---

31. Delhi 21, *op. cit.*, p. 25.
32. R.K. Verma (2002). "Straight Question," *Hindustan Times*, March 29, Delhi Edition, p. 8.

personnel familiar with the 'one window' approach could record all relevant data, while extending full assistance to the relocated industry owner.[33]

Under the Hazardous Waste Rules, 1989, the State governments were required to maintain inventories of disposal sites but this was not done in Delhi. Post-Rio grants ranging from Rs. 5 lakh to Rs. 15 lakh were made available to each State to identify and develop a secure landfill site for long-term containment. By 2002, three sites in Gujarat and one site in Andhra Pradesh had been identified. In the case of Delhi, a site was identified at Bawana but it could not be developed into a secured landfill because of public opposition. According to the HPC, Delhi government was among those State governments, which did not take up the development of disposal sites with any degree of seriousness. The absence of a site meant that the DPCC could not give directions about proper disposal which meant legitimisation of discharge of hazardous waste into open spaces and drains or into municipal land fill.[34] All three posed a danger to the environment and human beings.

Given the small scale and scattered nature of the sources of hazardous waste generation, Common Effluent Treatment Plant (CETP), as Delhi-21 had suggested was probably the best option. Delhi government had reason to be satisfied for securing funds for time bound construction of as many as 15 plants. Environmentalist Ravi Agarwal was sceptical whether Rs. 254 crore worth of investments in CETPs could substantially reduce the toxic waste generated every day in Delhi. He believed that the CETPs were being regarded as a 'magic solution', which in fact he said they were not. Of five functioning plants investigated by the Delhi-based NGO, Toxics Link, with which he was associated, not one was found to be working satisfactorily. CETPs could only be one link in a hierarchy of measures but that was not even being proposed. The industrial wastewater entering the CETP ideally needed to be pre-treated. If the effluent entering the CETP was very toxic, the CETP alone could not provide the necessary detoxification. Secondly, CETPs could handle only a range of chemicals and not 'a mixed broth.' Thirdly, the CETPs generated sludge or solid waste containing concentrated toxins that

33. Delhi 21, *op.cit.* p. 25.

34. *High Power Committee Report on Management of Hazardous Waste, op.cit.*

needed further containment or treatment for which there was no provision. In case of power outages, CETPs would be bypassed and untreated effluents would continue to be disposed of into vacant land or into drains.[35]

Apart from the formidable technological problems in the proper use of CETPs there were other obstacles in their efficacy which related to the rushed manner in which these had been constructed. In 2003, DSIDC acknowledged that there were many gaps in the data of NEERI about nature and size of operations, nature of available pre-treatment facility *etc.* on the basis of which CETPs had been designed.[36] Additional difficulties in the management of the plants were unavoidable due to the imposition of the dual authority system imposed by the Effluent Treatment Plant Act, 2000. As per the Act, management of a CETP had to be entrusted to a society of the users in the industrial estate. However, the society could not set the dues or recover unpaid dues from the individual users. The Act provided for the creation of an appropriate authority for deciding cost apportionment, recovering unpaid dues from the users and imposing penalties on the defaulters.[37] Till April 2004, only four CETPs had become partly functional though ten had been constructed. In 2003, the DSIDC issued notices to 17,000 industrial units asking them to pay an additional Rs. 73.8 crore towards the revised cost of installing the CETPs. Many industrialists refused to pay on the grounds that they had paid what the Supreme Court had ordered. Some had moved the Court.[38] Till the dispute was resolved, it could be assumed that unprotected public land and the Yamuna into which drains lead would have to bear the brunt of Delhi's industrial pollution.

It was difficult to be optimistic about sustained improvements in the management of hazardous waste in Delhi in the near future and not echo the sentiment of the HPC which had called the situation in India in relation to hazardous waste as 'grim.'

---

35. Agarwal, Ravi (2002). "Slowly, Slowly, Down the Stream," May 17, *Indian Express*, Delhi Edition.
36. Parashar, Vikas (2004). "CETP Standoff," *Down to Earth*, April 30, p. 17.
37. Down to Earth (2003). "For a Good Churn," *Down to Earth*, 15 January, p. 38.
38. Parashar, Vikas (2004). "CETP Standoff," *Down to Earth*, April 30, p. 17.

BIOMEDICAL WASTE

Nature of Biomedical Waste

Biomedical waste is the waste that is produced by hospitals, clinics and other medical and research facilities. It includes infectious, hazardous, radioactive and other general waste. Biomedical waste has been defined in the Biomedical Waste (Management and Handling) Rules, 1998 as any waste, which is generated during the diagnosis, treatment or immunisation of human beings or animals or in research activities pertaining thereto or in the production or testing of biologicals. The Rules define the biomedical waste more narrowly as it can encompass, human tissues, organs, body parts, animal tissues, organs, body parts, needles, syringes, scalpels, blades, glass, waste generated from disposable items other than the waste sharp such as tubing, catheters, intravenous sets, incineration ash and discarded medicines and Cytotoxic drugs.[39] In low income countries, biomedical waste is often got rid of by treating it as if it were household or office waste. Like industrial waste, biomedical waste constitutes a very small component of MSW. However, its distinct hazardous character involves specific social and environmental consequences. If its presence is ignored, there can be avoidable disruptions in the routine collection, transportation and disposal of MSW. Also, adverse impacts on health and environment, mitigation of which is the rationale of solid waste management, will not get the attention these deserve. Quest for a durable arrangement for safe disposal of MSW cannot be de-linked from a similar quest for biomedical waste.

Problem of Biomedical Waste in Delhi

Delhi like other major cities of India faces a grim situation in the handling of biomedial waste. The rapid increase in population has led to a sharp increase in the number of health-care facilities in Delhi. It is estimated that there are more than 3000 health-care facilities in Delhi of which more than 1200 treat/service 1,000 and more patients every month. The view of the Ministry of Environment and Forest, Central government, in 2000 was that health-care facilities in Delhi

---

39. Centre for Environmental Law Education, Research and Advocacy, *op.cit.*

generate between 20-25 tonnes of medical waste daily.[40] Since Delhi generates an estimated 7,000 tonnes of municipal waste per day, biomedical waste forms less than 0.5 per cent of Delhi's total waste. As per World Health Organisation (WHO) estimates, 85 per cent of the medical waste generated by health-care facilities is non-infectious, 10 per cent is infectious and 5 per cent is hazardous.

In Delhi, like in other cities of the developing world that have followed the footsteps of the West in matters of production-consumption and waste disposal, for biomedical waste, incineration has been the accepted solution for destroying the pathogen and reducing its bulk before the residue is dispatched to the landfill. However, an incinerator is a sophisticated machine which requires a regular supply of electricity as also a minimum load of waste to become functional. It requires maintenance and timely repairs which could be costly and a low priority matter for even big hospitals.

The condition of disposal of biomedical waste in hospitals in Delhi is very unsatisfactory. In February 1998, Srishti conducted a survey over a period of three weeks mainly during the morning hours, since it was then that maximum waste collection and disposal takes place. Some of the findings of the survey were:

- Of the hospitals with incinerators, only 14.2 per cent of the incinerators were operating at temperatures prescribed by CPCB.
- 50 per cent of the hospitals surveyed had given their waste disposal work to a contractor. Also 33.3 per cent of the hospitals had their incinerators run by contractors.
- Ram Manohar Lohia hospital was the only hospital that had a shredder installed for plastic waste.
- Plastic material from the hospitals were being collected for recycling.
- Many incinerator operators and waste collectors had no idea of the hazards of waste nor were they given any protective gear.
- A lot of infectious waste was being directed towards bins even in hospitals with incinerators.[41]

---

40. State of Environment Report for Delhi 2001, *op. cit.*, p. 65.
41. Agarwal, Ravi (1998). "Medical Waste, Issues, Practices and Policies," Srishti, p. 33, New Delhi.

Thus, it can be safely said that till recently, in Delhi, biomedical waste was being disposed of by and large with lack of awareness of its potential for harm to human health and the environment and deference was shown only due to its market value to the high grade plastic and metal components.

Social and Environmental Implications

Ensuring disposal of biomedical waste so that there is minimal adverse impact on human health and the environment poses a daunting challenge to city authorities of developing countries. "Medical waste mismanagement is not only linked with the spread of infectious agents like, Human Immuno deficiency Virus (HIV) and HBV, but also with indirect problems like endocrine disruption, cancers, reproductive disorders, immune suppression, nervous disorders etc. Equipments and chemicals used in hospitals, like mercury-containing instruments, radioactive isotopes, glutaraldehyde, Cytotoxic drugs etc. pose a big threat if they are not disposed of properly."[42]

The groups most vulnerable are the ward boys, nurses, municipal workers and ragpickers. Sharps, which include syringes' needles, have the highest disease transmission potential among all categories of medical waste. In recent times, diseases like HIV/AIDS virus and Hepatitis B and C virus have been shown to be transmitted considerably due to syringes containing contaminated blood. A WHO Report indicates that a hepatitis B virus can survive in a syringe for eight days.

When medical waste is incinerated—till recently the most widely used method to dispose of medical waste-pathogens or biological hazards are replaced by chemical hazards that are extremely expensive to monitor and control. The different types of toxins produced by medical waste on incineration are acid gases, dioxins and furans and heavy metals such as mercury, lead and cadmium.

Acid gases include nitrogen oxides that have been shown to cause acid rain formation and affect the respiratory and cardiovascular system. As plastic is incinerated, hydrochloric acid is produced. The

42. The Problem of Medical Waste. Toxics Link_files/issues.

acid attacks the respiratory system, skin, eyes and lungs with side effects such as coughing, nausea and vomiting.

Dioxins and organic chlorines are formed as a result of the combination of chlorine molecules in plastic catheters syringes, PVC tubing, bleached paper *etc.* with organic materials. Organic chlorines do not biodegrade. They bio-accumulate and are magnified up the food chain and end up in dairy foods and meat. They are proven carcinogens and endocrine disrupters and also weaken the immune system and damage the male and female reproductive organs.

Heavy metals are released during the incineration of medical waste. Their concentration increases in the residual ash in the incinerator if the filter can catch most of the pollutants and prevent their escape into the air. Mercury gets vaporised. It settles in water bodies from where *via* homogenisation in the food chain and bio-accumulation it reaches humans when they eat fish. It is only recently that it has become known that traces of mercury can affect the central nervous system, kidneys and liver and also stall ability to feel, touch and move. Cadmium and lead, present in plastic, accumulate in the incinerator ash from where when carted to the landfill, these substances can leach into the soil and groundwater. Chronic exposure to lead is associated with decreased intelligence and impaired neuro-behavioral development in children. Cadmium is a carcinogen linked with toxic affects on reproductive development, liver and central nervous system.[43]

Government of India Initiatives

Like the Municipal Solid Waste (Management and Handling) Rules, 2000, the Biomedical Waste (Management and Handling) Rules 1998 were substantially a product of interaction between concerned citizens, the Supreme Court and Government of India acting through the Ministry of Environment and Forest.

In 1995, the MoEF issued a draft rule requiring that all 50 bed hospitals install on site incinerators. After B.L. Wadera, who had earlier filed a PIL for directions to the MCD and NDMC for garbage removal, filed a PIL seeking directions to Delhi hospitals for enforcing

43. Nazareth, Sameer and Megh Kela Rathi (2002). "Medical Waste Incineration," in Srishti, (2002) *Hospital Waste-Time to Act*, Srishti, pp. 12-13, New Delhi.

this Rule, the Court issued the necessary order to the hospitals. Srishti and some other NGOs filed a counter petition in which they sensitised the Court and government agencies about contemporary views on the highly toxic nature of incinerator emissions. They were able to prevail on the MoEF. Both their suggestions about alternate safer technologies and the focus on standards were accepted.[44]

The Government of India notified Biomedical Waste (Management and Handling) Rules in 1998. These Rules, as amended in 2000, apply to all persons that generate, collect, receive, store, transport, treat, dispose or handle biomedical waste in any form. The 1998 Rules required the government of every State and Union Territory to establish a prescribed authority with such members as may be specified for granting authorisation and implementing these rules. This provision was amended in 2000 by nominating the SPCB/ Pollution Control Committees as the prescribed authority for the enforcement of these rules.[45] The penalty for violation of rules could be, as provided in the Environment Protection Act, 1986, a fine of Rs. One lakh and imprisonment for five years. Some of the emphases in the rules needs to be noted.

- Rule 4 states that "It shall be the duty of every occupier of an institution generating biomedical waste which includes a hospital, nursing home, clinic, dispensary, veterinary institution, animal house, pathological laboratory, blood bank by whatever name called, to take all steps to ensure that such waste is handled without any adverse effect to human health and the environment."
- According to Rule 5, "Biomedical waste shall be treated and disposed of in accordance with Schedule I, and in compliance with the standards prescribed in Schedule V," and "every occupier, where required, shall set up in accordance with the time-schedule in Schedule VI, requisite biomedical waste treatment facilities like incinerator, autoclave, microwave system for the treatment of waste, or, ensure requisite treatment of waste at a common waste treatment facility or any other waste treatment facility.

44. Ravi Agarwal, *op.cit.*, p. 30.

45. Centre for Environmental Law Education, Research and Advocacy, *op.cit.*

- As per Rule 6, "biomedical waste shall not be mixed with other wastes," and it "shall be segregated into containers/bags at the point of generation" in accordance with prescribed schedules.
- Rule 8 states that medical facilities providing services to 1,000 or more patients per month shall obtain authorisation from the prescribed authority in the context of precise arrangements he may make for handling the waste.
- According to Rule 10, "every occupier/operator shall submit an annual report to the prescribed authority in Form 11 by 31 January every year, to include information about the categories and quantities of biomedical wastes handled during the preceding year. The prescribed authority shall send this information in a compiled form to the CPCB by 31 March every year."
- According to Rule 11, "every authorised person shall maintain records related to the generation, collection, reception, storage, transportation, treatment, disposal and/or any form of handling of biomedical waste in accordance with these rules and any guidelines issued" and all such "records shall be subject to inspection and verification by the prescribed authority at any time."
- Rule 14 was added in 2000 to reinforce the responsibility of the local civic bodies to provide suitable common disposal/incineration sites for the biomedical wastes generated in the area under their jurisdiction. The new Rule provided that, "The Municipal Corporations, Municipal Boards or Urban Local Bodies shall be responsible for providing suitable common disposal/incineration sites for the biomedical waste generated in the area under their jurisdiction." In areas outside their jurisdiction, "It shall be the responsibility of the occupier generating biomedical waste/operator of a biomedical waste treatment facility to arrange for suitable sites" to comply with the provisions of these rules.

The basic principle enunciated in the rules is that biomedical waste be segregated and various components disposed of separately by the generators of the waste. Schedule I of the Rules indicates ten categories of medical waste and requires that these be stored in encoded coloured bags.

- Human Anatomical Waste (human tissues, organs, body parts), Animal Waste (animal tissues, organs, body parts, carcasses, blood and experimental animals used in research etc.), Microbiology and Biotechnology Wastes (waste from laboratory cultures, stocks or specimens of micro-organism, live or attenuated vaccines, human and animal cell culture used in research etc.) and Soiled Waste (items contaminated with blood and body fluids including cotton, dressings, soiled plaster casts etc.) are meant to be put in a yellow bag. Treatment and disposal options in respect of Human Anatomical Waste and Animal Waste are incineration and deep burial, an option available only in towns with population less than five lakh and in rural areas. Treatment and disposal options in respect of Microbiology and Biotechnology Wastes and Soiled Waste are in addition autoclaving and microwaving.
- Solid waste (waste generated from disposable items other than the waste sharp such as tubing, catheters, intravenous sets) is required to be put in a red bag. Treatment and disposal options are disinfection by chemical treatment, autoclaving, microwaving and mutilation and shredding. After disinfection, these can be shredded and made available for authorised reuse. No chlorinated or chemical waste can be incinerated.
- For waste sharps (needles, syringes, scalpels, blades, glass etc.), a puncture proof blue/white bag is recommended. These too can be disinfected which means an hour long soak in a one per cent hydrochloride solution or any other equipment chemical reagent and made available for plastic, metal and glass recycling after obstruction of reuse possibility or subjected to autoclaving or steam sterilisation between 121 and 150° C for 30 to 60 minutes.
- About liquid waste (waste generated from laboratory and washing, cleaning, house-keeping and disinfection activities) and chemical waste (chemicals used in production of biologicals, chemicals used in disinfection, as insectides, etc.), the Rules require that these should be disinfected by chemical treatment and liquids discharged into drains and solids in secured landfill.

- For the last two waste categories, incineration ash and discarded medicines and cytotoxic drugs, the requirement is that these be placed in a black bag that should be eventually sealed for dispatch to a secure landfill. For discarded medicines and Cytotoxic drugs, the option of incineration is also provided.

Delhi-21 Perspective

Delhi-21's view was that most of the 20 to 25 tonnes of biomedical waste generated every day found its way into municipal receptacles. In order to protect the environment and health of the general public a strategy of sustainable bio-waste management should have three objectives. Units generating biomedical waste should separate contaminated waste at source for collection and external treatment. There should be development of appropriate treatment and disposal sites. Finally, there should be emphasis on building awareness about sound biomedical waste principles and imparting training to concerned staff and operators about their application. Under the existing Rules it was the responsibility of biomedical waste generators to store, transport and dispose of the waste. Since many were too small to operate their own treatment facilities, the Department of Health Services (DoHS) or the municipal authorities could license private contractors to set up treatment facilities on a Build, Operate and Transport (BOT) basis. Delhi government or the appropriate municipal authority could provide the land and DPCC could monitor and regulate the activities of the private contractor.[46]

Interventions of Public Authorities 1998-2003

Efforts of Delhi Government

As required under the Biomedical Waste (BMW) (Management and Handling) Rules, 1998 as amended in 2000, the Delhi government issued a notification on 6 July 1999 nominating the Delhi Pollution Control Committee (DPCC) as the prescribed authority for enforcement of the provisions of these rules. The Delhi government also constituted an advisory committee to advise the government and the prescribed authority about matters related to the implementation

46. Delhi-21, *op.cit*, pp. 52,53.

of these rules. The notification also required every occupier generating biomedical waste to install an appropriate facility in his premises or set up a common facility to ensure requisite treatment of waste by 30 June 2000. Till December 2003 DPCC had granted provisional authorisation to 1,000 health-care facilities.

In keeping with the assessment that a large number of small clinics and private hospitals in the National Capital Territory did not have the capacity to set up and operate individual treatment facilities, eight autoclaves and shredders procured under the Japanese Debt Relief Grant Assistance were installed in eight Delhi government hospitals. The excess capacity of these was made available to nearly 1,500 major generators of medical waste to enable them to safely dispose of their waste. Until December 2003, there were 28 incinerators, 18 autoclaves, and two microwaves in operation.[47]

The services of the Centre of Occupational and Environmental Health at LNJP hospital were used to train 3,900 health care workers for implementation of the provisions of the Biomedical Waste Management and Handling Rules.[48] NGOs like Srishti and Centre of Environment Education were also mobilised to train health care workers. DPCC visited 44 major hospitals along with Srishti and Centre of Occupational and Environmental Health to review and assess the waste management practices of the hospitals.[49] Delhi government also constituted an internal inspection committee comprising of senior officials and experts who carried out surprise checks of the hospitals and advised them on the spot.

The Indian Institute of Technology, Delhi was awarded the work of testing the effluents and monitoring the incinerators in all the 100-bed hospitals.[50] DPCC granted authorisation to three private operators of facilities for collection, reception, storage, transportation, treatment and disposal of biomedical waste. Subsequently, it withdrew its authorisation from one operator. A new government facilitated operator is reportedly coming up in Delhi.[51]

---

47. Delhi Pollution Control Committee, *dpcc.delhigovt.nic.in*

48. *4 Years*, Working Report, *op. cit.*, p. 17.

49. Delhi Pollution Control Committee, *op.cit.*

50. *Environment*, (November 2002), Delhi Government, p. 4.

51. Delhi Pollution Control Committee, *op.cit.*

Assessment

In Delhi an enlightened perspective was in place with the notification of the Biomedical Waste (Management and Handling) Rules, 1998 in relation to a significant aspect of solid waste. The Rules favoured clean technology, waste segregation and a disposal strategy compatible with each waste stream. The new perspective was sought to be integrated with ground realities through appropriate interventions by the Delhi government during 1998-2003. Though these seemed to have made very little difference in the way public hospitals were handling their waste and additional steps which needed to be taken to consolidate the new paradigm could be identified, the outlook for the future in Delhi, in terms of reduction of adverse impacts on health and environment from hazardous waste from hospitals was promising.

The most significant feature of the 1998 BMW Rules was that by breaking sharply with the incineration option and restricting it to only six cases, the Rules provided for pre-emptive action to avoid a possible health crisis of which the situation in Bhopal, where 20 years later after the methyl iso cynate gas leak, 500,000 persons were chronically ill represented a sample. The chemicals that contributed to their condition are still not fully understood.[52] In 1998, disillusionment with incineration was rare. It was only in September 1997 that Environment Protection Agency (EPA) in the United States made the emission levels stringent reflecting the considered view that they were the predominant source of dioxins which is produced in the burning of plastic.[53] Dioxins are so toxic that their presence is measured in not parts per million, but in parts per billion.[54] The expectation of EPA was that 93 per cent to 100 per cent of small incinerators burning about 200 kg of medical waste per hour would shut down since it would be too expensive to retrofit them.[55] It was logical for their manufacturers to migrate to developing countries where standards were low and enforcement weak to hawk their wares. In a liberalising economy where people are in awe of foreign brands, awareness of

52. Down to Earth (2003), "Nurturing Bhopal", *Down to Earth,* December 15.

53. Ravi Agarwal, *op.cit.*, p. 5.

54. Brown, R. Lester, (2001), *Eco-Economy*, Orient Longman Limited, Hyderabad, p. 134.

55. Ravi Agarwal, *op.cit.*, p. 53.

BMW Rules could become a tool for helping health centre owners to choose wisely and promote public health while selecting a device for quick waste disposal.

As alternatives to incineration, the BMW Rules support autoclaving and microwave disinfection. Autoclaving is typically used for reusable items such as syringes and medical equipment. In autoclaving or steam sterilisation, medical waste is heated at 120 and 165 degree Celsius for 30 minutes to 90 minutes in chambers in which pressurised steam is introduced. The steam penetration ensures destruction of bacteria and pathogenic micro-organisms. Waste is reduced by an estimated 75 per cent of its volume and it can be land filled directly or compacted further. In hospitals in industrialised countries, up to 45 per cent of infectious medical equipment becomes re-usable through autoclaving. Autoclaves are commercially available in all sizes from desktop to industrial unit. Microwave disinfection relies on treating medical waste with moist heat and conventional microwaves. High frequency waves cause the molecules within the waste material to vibrate. This generates heat from within the material itself. The heat generated is high enough to ensure that all microbes are killed. Microwave is also economically competitive and comes in small and large sizes. Since the internal heating system is closed and has virtually no emissions, it has no need of any pollution control equipment.[56]

Changes in the BMW Rules made in 2000 ensured that state-of-art technologies which are cheaper and less material intensive would find a platform to demonstrate their unique aspects and find an opportunity to develop their potential. This provision could help in promoting a device which uses solar energy for disinfection of biomedical waste. Such an instrument was invented by a doctor from Indore and selected by CPCB for field trials.[57]

By stating categorically that chlorinated plastics could not be burnt even as plastic waste was growing exponentially—120 million syringes are generated every year due to immunisation programmes alone—the Rules were able to confer legitimacy on ecologically sound but old

---

56. Kela, Megha *et al.*, (2000). "Managing Hospital Waste—A Guide for Health Care Facilities" Srishti, p. 45-47, New Delhi.

57. Jamwal, Nidhi (2004). "Solar Treatment", *Down to Earth*, p. 26, March 15.

fashioned practices such a sterilisation. In 2003, a spokesman of Ministry of Health and Family Welfare, Government of India defended a decision of the Government of Haryana to rely on sterilisation in preference to plastic disposable syringes.[58] This augures well for the future.

In order to promote the new Rules, Delhi government chose to rely mainly on NGOs, who were familiar with the hospital environment, and asked them to create publicity material and training modules, for members of the staff in select public hospitals in Delhi such as the All India Institute of Medical Sciences (AIIMS). Their efforts may not have had much visibility because ideally management of all types of wastes in an institution should be regarded as part of the same problem—and MSW and human waste were not on the programme for which the NGOs had been mobilised. The exposure to complex ground realities enabled the NGOs to generate case studies, hand books, manuals on both technical and non-technical matters and fact sheets tailored to the needs of small health care centres, the fastest growing model among health facilities. The view of Megha Kela and the team from Srishti was that, "Training was a vital step in the process of implementing a waste management system. In this step, existing mind sets are changed or broadened."[59] The inspiration for Anu Goel, author of 'Small Price Big Returns' which discusses costs of bins, bags, needle destroyers and chemical disinfectants in addition to waste management formats in health care centres was the fact that, "During our work in setting up waste management programmes in hospitals and during seminars and meetings, we realised that budgets and investments are a major concern."[60]

Not lack of technology but absence of 'will' to apply it properly was the real problem in implementing the BMW Rules became clear from Srishti's fifth survey of medical disposal facilities in health-care units of Delhi in 2000. According to the Srishti survey, examination of 54 incinerators installed in health-care facilities revealed that 60 per cent of the incinerators were working in the range of 200-600 Centigrade and 20 per cent were operating in the range of 600-800 Centigrade and only 20 per cent of the incinerators were working in the specified

---

58. Jamwal, Nidhi (2003). "Prickly Issue," *Down to Earth*, July 31, p. 44.
59. Kela, Megha, Sameer Nazareth, Anu Goel, Ravi Agarwal (2000). *Managing Hospital Waste*, *op. cit* p. 30.
60. Goel, Anu (2002). "Small Price-Big Returns," *Hospital Waste*, June, *op. cit* p. 5.

temperature range of 800-1000 Centigrade. In addition, 38 per cent of the hospitals incinerated plastic waste was being disposed of along with other wastes. Further, none of the eight autoclaves installed in Delhi were operational. Proper sharps waste management practices were found adopted in only 60 per cent of the health-care facilities.[61] The last finding is also supported by the NGO Vatavaran's study which revealed that out of 40,000 syringes used every day only 40 per cent were destroyed and 60 per cent were reused.[62] Clearly, man-management and human resource development issues had been neglected in the new waste management initiatives. What was needed was a demonstrated commitment by the head of the institution to the new regime which involved creation of a cell under a senior functionary who had the means to ensure that all categories of personnel had the equipment, access to designated spaces and training to perform their role. Ravi Agarwal's view is setting up of waste management committee with representative of all sections of hospital workers is essential so that the waste related initiative does not become an un-wanted burden.[63]

However, technological issues were also sometimes the problem in the implementation of the Rules. Srishti's Fifth survey also showed that some hospitals had stopped segregating biomedical waste because of lack of access to a centralised facility for its treatment. Delhi Government had facilitated availability of three common disposal sites but many more were needed if on-site incineration was to be discouraged. The failure of Delhi government to develop a site for the safe storage of hazardous waste meant that residues from bio-medical waste were being dispatched to existing land fills.

Even three years after their notification, most government dispensaries in the city, whether under the MCD, NDMC or the Central government, each attending to over 1,000 patients seemed quite unaware of the BMW Management Rules and their provisions relating to source segregation or of the huge penalty in violating these Rules. In July 2001, a doctor working in a dispensary in Lodi Road had no hesitation in admitting that, "We generally throw the used

61. Fifth Srishti Survey of Medical Waste Disposal Practices in Health-care Units of Delhi, Quoted in *State of Environment Report for Delhi 2001*, p. 65.
62. Jamwal, Nidhi (2003). "Prickly Issue", *Down to Earth*, July 31, p. 44.
63. Ravi Agarwal, *op. cit.* p. 51.

syringes, cotton swabs, waste dressings, bandages, expired medicines etc. into the dustbin which is cleared by the sweeper."[64] Clearly, the concerned authorities had abdicated their responsibilities in not informing the health-care facilities about the Rules and ensuring their compliance.

Further, about 1,200 health care establishments had applied for authorisation under the Rules to the prescribed authority. About 1000 authorisations were issued and about 230 were denied authorisation and these included Central government hospitals such as Safdarjang Hospital and Employees State Insurance Hospital, Rohni, Army Hospitals such as the Army Hospital and the Base Hospital, Delhi government hospitals such as Hindu Rao Hospital and MVID Hospital and besides private hospitals such as the Escorts Heart Institute. The list of health care establishments which are operating without authorisation also include many Delhi government dispensaries.[65] There was need for the Delhi government through its Health department to ensure that at least the government hospitals followed the provisions of the Biomedical (Management and Handling) Rules and thus acted as role models for other medical establishments.

The DPCC could have assumed a more pro-active role in ensuring compliance to the Rules. It could have collected, compiled and published technical and statistical data relating to biomedical waste situation in the health care facilities and the measures devised for their effective control or abatement. In order that it could discharge these responsibilities, its organisational strength needed to be enhanced.

Thus, between 1998 and 2003, public authorities in Delhi were able to finalise and institutionalise a paradigm for the scientific and practicable disposal of biomedical waste. True, their effort involved only a fraction of MSW, but since the paradigm emphasised the holistic view of waste and waste management principles of reduce, segregation and reuse, prioritised human safety and environment protection particularly from the burning of plastic, it can be said that during 1998-2003, the cause of SSWM in Delhi had been advanced through the cooperative actions of Delhi's decision makers.

---

64. Vibha Sharma (2001). "Govt Dispensaries Fail to Meet Waste Disposal Norms", *Hindustan Times*, Delhi Edition, July 9.

65. *Delhi Pollution Control Committee, op. cit.*

5

Human Waste

In Indian cities, though the amount of human waste that is co-disposed with municipal solid waste is small as compared to the garbage generated everyday, issues relating to its safe disposal need to be specially considered. As is the case with chemical and biomedical waste, the hazardous character of human waste as compared to non-hazardous MSW involves specific social and environmental impacts which require to be addressed. Human excreta poses a problem in the proper collection and transportation arrangements of MSW and can be a source of soil and water degradation on a scale that SWM is not geared to control but which nevertheless must be reckoned with by concerned government agencies and not merely by the municipality since the whole rationale of SWM is protection of public health and maintenance of environment quality matters of abiding concern for all whether they are governors or the governed.

Problem of Human Waste in Delhi

An adequate sanitation system is generally regarded as one in which the toilet facility is connected to a sewer network linked to a few centralised wastewater treatment plants. This requires an expensive and complex material and water intensive infrastructure but many modern governments in low income countries consider its expansion the most practical method to meet a basic need of their citizens.

Sanitation facilities in Delhi have not been able to keep pace with the phenomenal increase in its population which has been due to the migration from the surrounding rural areas. The high economic return in the small-scale manufacturing sector in Delhi has had a magnetic impact on young men and women (men in the 15-39 age group, women in the 20-34 age group) from the rural areas of Bihar, Haryana, Punjab and Rajasthan. About three lakh new migrants come into Delhi every year in search of employment opportunities that are easily available in the unorganised and informal sector of Delhi's vibrant economy. Most of the migrants are illiterate and can find a place to live only in slum-clusters. In Delhi the density of population is 9294 per kilometre, which is the highest in the country. The decennial population growth rate between 1991-2001 was 46.31 per cent, which is also an indicator of high rate of increase of population and pressure on civic services in Delhi.[1]

The massive increase in population has led to encroachment of public land, unbridled growth of *jhuggies*, creation of slums and unauthorised colonies. According to the Economic Survey of Delhi 2001-2002, more than 47 lakh people lived in approximately six lakh *Jhuggie-Jhopris* in more than 1100 slum-clusters spread all over the city on encroachments mainly on public land. They also lived in Slum Designated Areas which were mostly concentrated in the walled city and its extensions, and which had been brought under the purview of the Slum Areas (Improvement and Clearance) Act of 1956.[2] About 3.5 lakh JJ dwellers lived in the 62,000 *Jhuggies* that had come up on the Yamuna riverbed and its embankments.[3] As per the policy of the Government of India, all notified Urban Slums and *Jhuggie-Jhopri* Clusters (JJC), without any exclusion, are considered eligible for provision of minimum basic civic amenities that includes toilets. The Slum and JJ Wing of the Municipal Corporation of Delhi provides sanitation facilities in the urban slums and JJ clusters directly or through NGOs. Since the sewer system is not expected to extend into these areas the toilets usually take the form of a common facility where flush toilets are connected to septic tanks or a water tight tank mounted on wheels.[4] Available sanitation facilities are also inadequate in other areas, where sewer extension is not legally excluded, but technological difficulty in laying pipes, or lack of financial resources on the part of the concerned governmental agency has prevented the expansion of sewage facilities. Political differences between the

1. *Tenth Five Year Plan of Delhi, Urban Development*, Planning Department, Government of Delhi, *www.delhiplanning.nic.in*, pp. 1420 and 1422.
2. *Economic Survey of Delhi, 2001-2002, Chapter 14, Urban Development*, Planning Department, Government of Delhi, *www.delhiplanning.nic.in*
3. *Yamuna Action Plan, www.yap.nic.in*
4. Booth, Beverley, *et al.* (2001). *Urban Health and Development—A Practical Manual for Use in Developing Countries*, Macmillan, London, p. 274-275.

Central and Delhi government agencies involving payment of infrastructure costs can further stall the laying down of sewer and connection with the main trunk sewer and STP plants. More than 45 per cent of Delhi's population is living in settlements where a sewerage system is not provided.[5]

When the Congress Party formed the government in Delhi in 1998, the position of sewerage facilities in different types of unplanned settlements, in respect of which data is available, was as follows[6] (Table 5.1).

Table 5.1

Sewerage Facilities in Different Types of Settlements in Delhi (March, 1998)

Sl. No.	Type of Settlement	Total Number of Settlements	Sewerage Facilities 1997	Sewerage Facilities 1998
1	Urban Villages	126	80	83
2	Regularised Unauthorised Colonies	567	292	292
3	JJ Resettlement Colonies	44	22	24
4	Unauthorised Colonies	1017	None	None

Source: Delhi Jal Board.

Since no sewerage service is available in many habitats in Delhi, the sanitation related options for the residents of such areas are a private on-site or community on-site facility or open defecation. In densely populated colonies it may not be possible to accommodate watertight septic tanks which are also expensive to construct. The alternative is a small pit or a loose receptacle, a basket or a bucket[7] that has to be regularly emptied for which a spot on the public street or one near a waste collection centre is probably the most convenient place. It is significant that in the MCD's Preparatory *Bhagidari* Workshop for Solid Waste Management held in June 2003, Sanitary Inspectors, Assistant Sanitary Inspectors and Sanitary Guides listed open defecation as one of the sources of the problems they faced

5. *Delhi 21, Delhi Urban Environment and Infrastructure Improvement Project (DUEIIP) Report, (January 2001)*, Government of India and Government of National Capital Territory of Delhi, p. 43.

6. *Economic Survey of Delhi, 1999-2000, Chapter 13*, Water Supply and Sanitation, Planning Department, Government of Delhi, *www.delhiplanning.nic.in*

7. *Booth, Beverley, et al.* (2001). *op.cit.*, p. 275.

during their work of waste collection and counted disregard of their health concerns as one of the sources of their difficulties *vis-à-vis* the MCD.[8] Community toilets are not fully utilised due to a variety of reasons. Sulabh International, world renowned pioneers of innovative low cost sanitation committed to providing not only toilet pits/toilets but also bathing facilities at nominal cost to users in slum areas through an internal cross subsidy programme, manage many of the community toilets in Delhi. In practice, the community toilets are by-passed because they are considered to be too expensive, too far and too dirty.[9]

The problem of human waste disposal in Delhi has been aggravated by the poor condition of the sewerage system. The existing sewerage conveyance system is a large network of branch peripheral and trunk sewers approximately 560 km in length. There are 28 main trunk sewers which are about 130 km in length. The trunk sewers were laid at different time periods and some are between 40-60 years old. In the older ones the level of silting varies between 30 and 70 per cent and some sections have collapsed ('settled'). In the beginning of 2002, it was estimated that 91 km of trunk sewers required 'rehabilitation' which meant de-silting and wall repair. Since sewage could not flow into the sewage treatment plants large quantities of human waste was diverted into the storm water drainage system.[10]

Social and Environmental Implications

The National Commission on Urbanisation, has pointed out that the urban centres in the country generated the most brutal and inhuman living conditions, with large sections of the population (nearly 30-50 per cent in the case of Mumbai and Delhi) living in squatter settlements. "Overcrowding and the scarcity of water and sanitation in the slums lead to severe health problems and the abject degradation of human life."[11]

8. *Report on Preparatory Bhagidari Workshop for Solid Waste Management, 23-25 June 2003*, Asian Centre for Organisation Research and Development (ACORD), New Delhi, p. 16.
9. Ruet, Joel, *et al. Water and Sanitation Scenarios in Indian Metropolitan Cities—Resources and Management in Delhi, Calcutta, Chennai and Mumbai*, CSH Occasional Paper No. 6, 2002, French Research Institute, p. 142.
10. *Tenth Five Year Plan 2002–2007 and Annual Plan 2002-03* Vol. II, Planning Department, Government of Delhi, *www.delhiplanning.nic.in*, p. 1377.
11. *Economic Survey of Delhi, 1999-2000*, Chapter 14, Urban Development, Planning Department, Government of Delhi, *www.delhiplanning.nic.in*

Many health problems are linked to inadequate and poor sanitation. Human excreta is an extremely hazardous substance and the cause of most water-borne diseases. According to Hardoy, "At any one time, close to half of the urban population in Africa, Asia and Latin America is suffering from one or more of the main diseases associated with inadequate provision for water and sanitation." The key to sanitation is to 'ensure no human contact with excreta' and its safe disposal so that it 'does not contaminate water or allow the parasitic worms in excreta to survive, develop and re-infect humans.' Most of the diseases associated with contaminated soil and water including all water-borne diseases such as cholera, diarrhoeal diseases such as amoebiasis, giardiasis, E-coli, etc., enteric fevers such as paratyphoid and typhoid, hepatitis/infectious jaundice, enterobiasis (caused by pinworms), ascariasis (caused by roundworms) and schistosomiasis, hookworm and tapeworm, and filariasis are excreta-related. Nearly 50 per cent of these infections can be transferred from an infected person to a healthy person. The worst affected by these are the lower-income urban households. Hardoy says, "Removing and safely disposing of excreta and waste water is also a critical environmental health need yet many governments and international agencies have concentrated on improving provision of water and not on sanitation and on ensuring the safe removal of waste water," which is where the health hazards really lie.[12]

Absence of appropriate sanitation also affects physical safety and mental health. Hardoy *et al.* have cited many studies to show how as open and secluded spaces become scarcer, as cities grow and available spaces get used up for buildings, those who rely on open defecation, especially girls and women, have to face many difficulties. Women and girls need to wait until dark to avoid unwanted attention from men. A study in Bangalore showed "how women or girls face particularly severe problems of harassment and risk of rape when they defecate outside." Another study of a low-income settlement in Kanpur showed that "many did not use a community toilet because mothers did not have time to accompany their children and young girls to use it."[13] In yet another study, one woman reported that, due to lack of toilets,

12. Hardoy, Jorge E, Mitlin, Diana and Satterthwaite, David (2001). *Environmental Problems in an Urbanising World*, Earthscan Publications Ltd., London and Sterling, VA, pp. 39, 40, 57.

13. *Ibid*, p. 59.

"she used to feel giddy from suppressing herself" and that she would "try to drink less water," which affected her in other ways. Headaches were common among women. "The condition of those who are pregnant and ageing is particularly pitiable."[14] Besides the inconvenience, the main problem is the "humiliation of having to defecate with no privacy or protection."[15]

The commonly used alternative to a centralised sewer based system is a soak pit or septic tank which needs to be periodically cleaned. Ideally the receptacle should have remained closed for two years, in which case the excreta would have become free of disease causing organisms and transformed into an excellent soil conditioner. That, of course, is simply not practical. In the absence of suction machines to carry out periodic removal, human beings have to undertake this most hazardous, demeaning and disagreeable task. Since 1993, when the Employment of Municipal Scavengers and Construction of Dry Latrines (Prohibition) Act was enacted, it is illegal to engage anyone to perform this task. However, because of extreme poverty, some people are always available for such work. In Bangalore, a group of four persons gets about Rs. 400 for cleaning a septic pit.[16] The urban poor who come forward to do this work are least equipped to survive its destructive impacts.

The technological constraints of a sewer based system have imposed a huge burden on Delhi's environment. Wastewater in the sewers is meant to be treated for sharp reduction of its organic load at treatment plants before it is discharged into the river. In Delhi the main reason for the steady rise in the pollution in the 22 Km. stretch of the Yamuna which passes through Delhi is that this is not happening. Apart from the fact that sections of trunk sewers have deteriorated due to age and sewer lines have become blocked due to road construction work the capacity of the treatment plants has not been augmented to keep pace with the level of waste water which requires treatment. In the Ninth Five Year Plan which ended in March 2002 the targeted capacity of Sewage Treatment Plants (STPs) was 601.4 million gallons per day (MGD) against which the achievement

14. Sridhar, Lalitha, *Lack of Sanitation Causes Distress and Death*, *www.infochangeindia.org*
15. Booth, Beverley, *et al.* (2001). "Improvements in Sanitation," *op. cit.*, p. 267.
16. "Going to Pot," (2003). *Down to Earth*, June 15, p. 40.

was 482.4 MGD. A shortage of cement steel and cash was regarded as the main constraint.[17]

"Sewage systems are built to protect public health but badly managed sewers become a serious health hazard." If a sewer pipe is blocked, a connection is made with a storm water drain and untreated sewage outfalls not only pollute rivers but also contaminate groundwater and piped water supply systems through possible leaks. Pathogens infiltrate into drinking water pipelines, especially when these pipes do not have water, which is the case in many cities in developing countries, including Delhi, because there is no supply of water round the clock. "All this makes water-borne sewerage a waste disposal paradigm that is extremely expensive because of its high economic, environmental and public health costs."[18] According to the Planning Commission's assessment, 80 per cent of the pollution in the Indian rivers is due to human sewage.[19]

Government of India Initiatives

According to the Indian Constitution, sanitation is a state subject and therefore its provision is the primary responsibility of the state governments. The policy of the Central government since the 1970s has been to offer some assistance to urban centres to enable them to expand installation of sewer based waste disposal systems and make arrangements through alternate low cost technologies, if necessary, to meet the sanitation needs of the urban residents especially the poor. In view of the constraint of budgetary resources towards the end of the 1990s, it was considered unavoidable that the Urban Water Supply and Sewerage Schemes should increasingly depend on institutional finance and that state budgetary support made adequate to meet the counterpart matching requirements of the institutional finance. According to the Ministry of Environment and Forests, the Central government has taken various initiatives to improve the situation since the 1970s.[20]

---

17. *Tenth Five Year Plan op. cit.* p. 1376.
18. "Sewer Systems Totally Destroy Nature's Nutrient Cycle," (2002). *Down To Earth*, February 28, p. 30.
19. Nadkarni, Manoj (2002). "Drowning in Human Excreta," *Down To Earth*, February 28, p. 34.
20. *Agenda 21: An Assessment* (2002). "Urban Governance and Services," Chapter 15, Ministry of Environment and Forests, Government of India, *www.envfor.nic.in*

These included:

- Environmental Improvement of Urban Slums Scheme (EIUS). The scheme was initiated in 1974 and is applicable to notified slums in all urban areas. It aims at:
  - Provision of basic amenities such as water supply and sanitation.
  - The EIUS scheme was made an integral part of the Minimum Needs Programme in 1974.
- Integrated Development of Small and Medium Towns (IDSMT). This scheme was initiated in 1979 with a view to:
  - Augmenting civic services
  - Strengthening municipalities through promotion of resource generating schemes
  - Reducing migration from rural areas to larger cities by providing sufficient infrastructural facilities, including water supply.
- Urban Basic Services Scheme (UBSS) initiated in 1986 and Urban Basic Services for the Poor (UBSP) programme initiated in 1990-91. The primary objective of these schemes was to improve the standard of living of urban low-income households, particularly women and children, through the provision of sanitation and social services in slum areas. In 1990-91, the schemes were integrated with the EIUS and came to be known as the Urban Basic Services for the Poor (UBSP) programme.
- National Slum Development Programme (NSDP). This programme commenced in 1996 and under it additional Central assistance is being released to States/Union Territories for the development of urban slums. Objectives of the programme include:
  - Provision of adequate and satisfactory water supply, sanitation, shelter upgradation, garbage and solid waste management in slums.
  - Development of community infrastructure, empowerment of urban poor women and involvement of NGOs and other private institutions in slum development.

Besides the above schemes, the Central government also initiated two exclusive schemes for improving sanitation services in rural and urban areas. In respect of rural sanitation, a restructured Centrally Sponsored Scheme with the involvement of the local people and NGOs has been set in motion from 1993-94. In the case of urban sanitation, the limited focus under the ongoing Centrally Sponsored Scheme is limited to the welfare objective of eliminating the practice of scavenging through conversion of the existing dry latrines into sanitary latrines and rehabilitation of the liberated scavengers and their dependants.[21]

While the need for providing and improving water and sanitation facilities, which have been conceptualised as related activities, has been reiterated in the successive Five Year Plans of the Central government, the actual increase in the public outlay on these activities has been very modest and limited to specific objectives. The outlay rose from 1.4 per cent of the total outlay in the Eighth Five Year Plan to mere 2 per cent in the Ninth Five Year Plan.[22]

These programmes have yielded limited results in terms of the population covered by these services. Access to sanitation facilities is now available to 49.3 per cent of urban population (Ministry of Finance, 2000-2001).[23] About half of the population which has sanitation facilities are connected to sewerage and septic tanks and the balance half have low cost sanitation facilities. The coverage in rural areas is hardly 20 per cent.[24] The drive to do away with scavenging system in which the toilets not connected to sewers are manually emptied and cleaned got under way with the passing of the Employment of Manual Scavengers and Construction of Dry Latrines (Prohibition) Act, 1993 and up to March 2000 almost Rs. 1340 crore was spent on this scheme. "Yet less than 8 per cent of the recorded dry latrines were converted to sanitary ones till the first three years of the Ninth Plan."[25]

Prospects of safe disposal of human waste in urban areas were expected to improve due to the Central government schemes meant

21. *Annual Plan of India, 1999-2000*, "Social Development," p. 7, *www.planningcommission.nic.in*

22. *Agenda 21: An Assessment, op. cit.*

23. *Ibid.*

24. Pathak, Bindeshwar (2003). "Sustainable Sanitation Technologies: Sulabh Experience", *Seminar on Sustainable Development of Delhi: Emerging Challenges and Policy Options in the Era of Globalisation*, p. 3, June 16-17, Institute of Human Development, New Delhi.

25. Nadkarni, Manoj, *op.cit.*, p. 34.

to increase houses for the urban poor. Environment Improvement Slums Scheme (EISS) is reported to have benefited about 375 lakh urban slum-dwellers, against the targeted 330 lakh people. (Ninth Plan and MoUDPA, 2001). However, as the Central government itself acknowledged, despite the introduction of policies and programmes for providing housing to disadvantaged groups and efforts to contain the proliferation of slums, "the housing problem persists, resulting in pressure on urban services and infrastructure" (Planning Commission, 2000). Increasing population pressure on land and infrastructure and the associated high costs have made proper housing inaccessible to the poorer segments of the population and have caused a growth of slums and squatter settlements" (Eighth Plan). Therefore, in spite of a steady growth in housing stock, infrastructure and services, the demand-supply gap has kept on rising (Ninth Plan).[26]

Another major initiative of the Central government that impacted upon the prospect of human waste management in Delhi was the Yamuna Action Plan that was initiated in 1993 to clean the Yamuna River by the National River Conservation Directorate (NRCD), Ministry of Environment and Forest (MoEF) of the Government of India in 12 towns of Haryana, eight towns of Uttar Pradesh and in Delhi. Delhi was especially included because over the last two decades agricultural land that used to absorb sewage water had turned into residential colonies. Secondly, all drains that carried waste were outfalling into the Yamuna. Many storm water drains were in practice also being used as sewers. Lastly, over the years, lakh of people had made their homes on the riverbanks and the bed of the Yamuna and were treating the river as a receptacle of their wastes.[27]

Delhi-21 Perspective

The assessment in Delhi-21 was that "The authorities faced a daunting challenge in ensuring durable improvements. They not only had the task of "planning and catching up on the deficiency of appropriate sanitation or sewerage in the city but also planning and providing infrastructure for future urbanisation within the NCT." In an affirmation of the SSWM principles articulated at Rio, Delhi-21

26. *Agenda 21: An Assessment, op. cit.*

27. *Yamuna Action Plan*, yap.nic.in

said, "the long-term objectives must be to provide for all equal access to appropriate sanitation and drainage" so that "the environmental and health impacts of poor sanitation are greatly reduced" and the Yamuna river "meets its prescribed quality classification."[28]

For advancing towards this goal, there was need to extend the centralised water borne sewer and drainage net work to serve all areas, maintain it in good condition by "systematic repair and preventive maintenance," connect it to sewage plants which may have to be upgraded or newly constructed. There should be "local wastewater treatment where primary sewers are not available" and "appropriate sanitation where sewerage was not practical." Any expansion of sewage network must be in conjunction with the water supply network. Sewer design should facilitate easy access for inspection and de-silting.

About institutional and financial reforms that would be required, it was suggested that the Delhi Jal Board needed to increase its technical capacity and equipment from its present levels in order to become an efficient and responsive sewerage undertaking. It should have the autonomy to raise sewerage costs, which should include the cost of collection, treatment and operations. An independent regulator should ensure service delivery targets and rates that are affordable by consulting consumer associations. As in the case of municipal waste, sustainable management of human waste required "improved public relations, information and transparency of operations," and "consumer involvements at all levels from planning to implementation." NGOs must be encouraged to facilitate on-site sanitation of unsewered areas. To ease the financial burden on the government, private sector participation should be mobilised—for a start private developers should provide sewerage and treatment facilities which they should operate and maintain. Since the poor collection and indiscriminate disposal of solid waste in the sewerage and drainage system frequently caused blockages in the system, there was need to eliminate the solid waste from the sewerage and drainage system by simultaneously improving the solid waste management operations.[29]

---

28. *Delhi-21*, *op. cit.*, p. 44.

29. *Ibid*, p. 43.

Delhi-21 indicated a programme of investments required for improvement in management of human waste in Delhi in the short term (2001-2006), medium term (2007-2011) and the long term (2012-2021) to the extent of Rs. 46,000 million. In these projections, a meagre investment of Rs. 1,400 million or about 3 per cent of the total investment, was projected to facilitate on-site sanitation/ community toilets for the poor.[30] Absence of any substantial investment for this activity shows that tested low cost 'sewer-less' safe disposal alternate technologies were not considered a serious option for bringing about improvement in Delhi, even though it was felt that there was a "desperate need" to reverse the existing situation.

Interventions of Public Authorities (1998-2003)

Delhi Jal Board

In 1998 the Delhi government and the DJB had their work cut out for them in the matter of initiatives for a more sustainable method of managing human waste. According to an affidavit filed before the Supreme Court by the Ministry of Urban Development and Poverty Alleviation in 1994, these entities were committed to an Action Plan for abatement of pollution to the Yamuna. The projects included the refurbishing of the entire Trunk Sewerage system by December 2003, completion of the construction of five Sewage Treatment Plants (STPs) according to schedule, construction of additional STPs on drains where possible and implementation of guidelines to ensure that in the new STPs the Biochemical Oxygen Demand (BOD) level was 10mg/ litre rather then 20 mg/litre as was the case then.[31]

By March 2003 out of 567 regularsied unauthorised colonies sewer lines were laid in 458. Ninety eight out of 135 urban villages were similarly provided for. It was not considered technically feasible to do so in the case of at least 33 villages. By 2003 all 44 resettlement colonies had sewer lines.

Good progress seems to have made in the matter of sewage treatment facilities. By the end of 2003 there were 19 STPs including two newly constructed ones with 2.2 Million Gallon/Day (MGD)

30. *Ibid*, p. 45.

31. *Tenth Five Year Plan 2002-2007* and *Annual Plan 2002-2003*, Planning Department, Government of Delhi, Vol. II, *www.delhiplanning.nic.in*, p. 1377.

capacity at Delhi Gate *Nallah* and Dr. Sen Nursing Home *Nallah*. This had become possible because the Japanese Bank of International Cooperation (JBIC) had been persuaded to fund a second phase of the Yamuna Action Plan. However, in the crucial areas of desilting and rehabilitating sewer lines advance was slow. Only 20 km was covered by December 2003 and another 10.5 km stretch was expected to become adequately functional by June 2004.[32]

The Municipal Corporation of Delhi

Since the late-1980s, the scheme of construction of Jan Suvidha Complexes was pursued in Delhi to take care of the environmental problems caused by defecation in the open by the slum and *jhuggi* dwellers. The scheme, initiated in the Seventh Plan, was meant to cover the then 9.51 lakh slum residents staying in the notified slums and about 13 lakh residents staying in *Jhuggies*. It was estimated that the number of *Jhuggie* residents has now increased to about 30 lakh.[33]

During the Ninth Five Year Plan, 292 conventional/pre-fabricated Jan Suvidha Complexes, containing 7730 WC seats were made operational. 72 Mobile Toilet Vans, containing 1008 WC seats, were also added.[34] The scheme was able to touch only the fringe of the problem. According to Delhi government's own estimates, the number of JJ residents increased from 13 lakh in 1990 to 20 lakh in 1994 and were at present estimated to be 30 lakh. It can be assumed from these figures that since JJ residents are increasing at the rate of about 1.70 lakh per year, about eight lakh new JJ residents were added during the Ninth Five Year Plan period. The addition of 8738 WC seats during the same period meant that one WC seat was added for 91 persons as against the accepted national norm of one WC seats for 20-25 persons accepted under the scheme. The Delhi government has accepted that it was not possible for it to follow the national norms because of "physical constraints."[35] Against this background, the MCD, the agency responsible for meeting the growing demand for public conveniences,

32. *Annual Plan 2004-2005*, Vol IV and *Annual Plan 2001-02*, Planning Department, Government of Delhi. delhiplanning.nic.in, pp. 32-33, 36.

33. *Tenth Five Year Plan of Delhi*, Urban Development, *op. cit.*, p. 1423.

34. *Appraisal of Ninth Five Year Plan (1997-2002) and Annual Plan 2001-02*, Planning Department, Government of Delhi, *www.delhiplanning.nic.in*

35. *Tenth Five Year Plan of Delhi*, Urban Development, *op. cit.*, p. 1445.

decided to create 48 toilet blocks on a Build-Operate-Transfer basis in different locations. The entrepreneurs who had agreed to undertake the construction were expected to make a profit by charging a fee for the use of the toilets and leasing wall spaces to advertisers.[36]

Towards the end of 2001, construction work for yet another 956 Community Toilet Complexes (CTCs) involving 28,000 WC seats was in full swing. A combination of factors was responsible for this development. The Supreme Court of India admitted the public interest litigation filed by the Centre of Science and Environment (CSE), INTACH and Pani Morcha on the deterioration in the quality of the water of the Yamuna River during its passage through the Delhi Urban Agglomeration, despite the existence, since 1993, of the Yamuna Action Plan to sharply reduce pollution levels. The Supreme Court ordered the Delhi government to clean up the Yamuna River by 31st March, 2003. In 2001, the Ministry of Environment and Forests and Delhi government, as a part of their endeavour to comply with the Supreme Court's orders, made Rs. 166.4 crore available to MCD for construction of CTCs for slum rehabilitation etc. along with construction of Micro and Mini Sewage Plants and implementation of a Public participation and Awareness Programme under the Yamuna Action Plan (YAP). Funds were made available by the Japanese Bank of International Cooperation (JBIC) with a deadline on expenditure.[37]

The largest component of the work related to the construction of 1146 community toilet complexes. Due to non-availability of suitable sites, 956 CTCs were constructed under the programme at an estimated cost of Rs. 164 crore. From the comments of Shri S.K. Mahajan, it was clear that he recognised that there was a unique citizen, environment friendly and cost effective dimension to his new responsibilities which were actually inspired by the JBIC deadline and legal compulsion to reduce river pollution.

The CTCs built under the MCD's supervision were generally single or double storied, in the case of space constraint, with 20, 30, or 40

36. *Background Note on Conservancy & Sanitation Engineering Department of Municipal Corporation of Delhi.*

37. *Yamuna Action Plan, op. cit.*

WCs in each unit. On the lines of the Sulabh model that already existed in Delhi, there were facilities for both bath and toilet, separate sections for men and women with different entrances and toilet seats for children in the women's section. An additional feature was western style WCs for the handicapped. Maintenance of the CTCs was allotted for a three-year period to select NGOs and CBOs, who had first registered with the MCD, after accepting the latter's terms and conditions, and subsequently procured a license in an open auction. In earlier times, the established sanitation NGO, Sulabh International was given funds to construct CTCs on mutually agreed terms, which included exclusive responsibility for maintenance for the next 30 years. In situation of neglect and inconvenience to the public, MCD had been unable to intervene even as it was considered responsible for its functioning.

In the new system, as per the terms of the contract, the caretaker NGO was required to keep the facility open for 24 hours, provide a sanitation worker on the premises and ensure that there was a generator back up to take care of power outages. Permission for boring a tubewell was assured to the caretaker NGO by the MCD, eliminating dependence on the DJB for water. A minimum tariff of Re. 1 was fixed and the facility was made free for children below the age of twelve.

In the four resettlement colonies on the outskirts of Delhi, Molar Bund, Bakkarwala, Holambi and Tikri Khurd, squatter families from various locations inside Delhi were expected to be shifted in the near future. In the first two locations, 5,000 families were to be accommodated and in the other two locations 3,000 and 2,700 families were to have their homes. In the sewerage and the STPs to which the CTCs in these colonies were connected, a new decentralised simplified sewerage system, well tried in other countries such as Brazil which Delhi-21 had recommended for new group housing societies was established. Within these residential complexes, STPs of two and three MLD were constructed. Due to the FAB and SAFE technology that these were geared to apply, wastewater from the CTCs could be treated and made available for use in irrigation. The faecal coliform in the wastewater was expected to be 1000 MPN/100 ml of water which was within safety limits for horticulture use.

Japanese technology was also availed of in setting up 10 Micro Sewage Treatment Plants at various places in Delhi. Two were set up in Sector 15 of Dwarka, three in Hastsal, one in Anand Parbat and two each in Sultanpuri and Baljeet Nagar. It was expected that wastewater could be reused for flushing or for horticulture.

Since some of the CTCs constructed under the Yamuna Action Plan were connected to existing sewers and some were constructed with their own septic tanks, part of the funds were used for procuring jet-cum-suction machines along with ancillary equipments to remove the sludge from septic tanks and for cleaning blocked sewers.

Residual funds of Rs. 216 lakh were set aside for 'public participation' and awareness building. Community mobilisers were expected to attract people around CTCs for increasing its user profile. Multimedia training package were expected to be produced for caretakers of CTCs. Existing and newly constituted eco-clubs were to be used to generate groups that could visit facilities such as the Micro and Mini STPs and CTCs and the electrical crematorium that had been created to reduce the pollution in the Yamuna River. It was expected that these captive groups could also be shown how solid waste could be disposed of in a more environmentally benign fashion. Preparation of print and audio-visual publicity material on themes such as low-cost sanitation and waste management was also on the agenda. The funds were also meant to subsidise public events such as street plays, fairs, rallies, *shramdans* etc., where the message of a less polluting life style was to be reiterated.[38] By and large, funds were spent on the lines envisaged.

Assessment

In a city that was growing at the rate of 1000 persons a day, how adequate were the steps undertaken between 1998 and 2003 by the Delhi Government, Jal Board and MCD to ensure safe disposal of human waste in conditions of dignity and substantial reduction of possibility of diseases like jaundice and cholera through exposure to raw waste and consumption of contaminated water?

38. *Ibid.*

There was no basic change in the manner in which human waste was managed at the end of 2003 as compared to before 1998. This was clear from the fact that there was no improvement in the water quality of the Yamuna the final receptacle of the most of the 'grey water' and 'black water' that was generated in Delhi. In June 2003 a report in the Times of India noted that "even as the State government gears up for another three day 'symbolic' clean-up as a run up to the World Environment day on June 5, the Yamuna River is as polluted as ever. A ten year old clean up efforts started in 1993, seems to have literally gone down the drain." The BOD levels in the Yamuna it was noted, hovered between 15 and 30 micrograms per litre, while the normal BOD level should be just three mgs per litre. Higher BOD levels mean lesser chances of survival for plants, fish and other aquatic life in the river as the oxygen supply for organic matter was reduced. The coliform count varied between one lakh and 10 lakh, while the normal count should be at 5,000. Higher levels of coliform meant more risk of water-borne diseases.[39]

Sewage was the chief source of pollution of the Yamuna, as in the case of most other rivers in India. This was not adequately grasped by Delhi's governing authorities. In a one-day workshop in June 2002 organised by The Energy Research Institute (TERI) to evaluate the work done towards cleaning the Yamuna, Ms. Sheila Dikshit expressed the view that the major problem was that 10 per cent of Delhi's 600,000 *Jhuggis* were located on the banks and bed of the Yamuna and the slum dwellers did not have toilet facilities with 'running water.' The clear implication was that if they could be relocated and conventional toilets provided to them the problems could be substantially mitigated. Mr Baalu, Minister in the Environment and Forest Ministry expressed the view that there was a shortage of funds for pollution control measures. He was optimistic about his negotiations with the Japanese Bank of International Cooperation for a grant for a second phase of the Yamuna Action Plan. He hoped that Delhi government would share the burden on a 30-70 basis.[40]

What seems to have been glossed over was the fact that regardless of where the slum dwellers were shifted and the quantum

39. Roy, Subhajit (2003). "10 years on, Yamuna Still Toxic," *Times News Network*, June 2.
40. The Energy Research Institute, *www.teriin.org/news/jun023.htm*

of additional money spent on toilets unless there was an alternate viable system for final human waste disposal which by-passed the river, the problem of river pollution would persist. Further, considering the costs involved, it was highly unlikely that Delhi government would be able to secure funds to pay for deficiencies in existing sewer based sanitation infrastructure while providing for its expansion to meet further needs. In the second Yamuna Action Plan, 94 crore was allocated to complete the repair work of just a single trunk Sewer – the one on/under Ring Road. 92 crore was set aside for the renovation of a single STP—the Keshopur STP—which had the capacity to treat only 72 MGD. A new STP, which could treat only 30 MGD was to be built at Okhla at a cost of 70 crore.[41] According to the study, carried out by M/s TCE consulting Engineers as much as 726.4 million gallons of sewer water was required to be made relatively pollution free every day by the STPs by 2006[42] to meet minimal public health requirements.

An issue of concern to all stakeholders of a sound sanitation system for Delhi— Delhi government, Delhi Jal Board, MCD and Government of India— was the shortage of water that Delhi was likely to face in the coming years due to limits in its augmentation from the neighbouring states. In the Annual Plan for 2004-2005, 45 lakh was provided to the Urban Development Ministry of Delhi for a Water Mission and promotion of new technologies for conservation and harvesting of rain water.[43] Since conventional sewer based systems are water intensive, the budget head could have included a survey within India and abroad of alternative waste water treatment systems as also of new kinds of toilets pioneered in Europe and in use in Asia and Africa, since 1997 which include those that are supposedly water-less, odourless, chemical free, light weight, easy to install and also environment friendly.[44]

DJB also took no particular interest in ensuring the success of public toilets constructed by the MCD. Shortage of water, which is

---

41. *Annual Plan 2004-2005*, Planning Department, Government of Delhi, Vol IV, *www.delhiplanning.nic.in*, pp. 33-34.
42. *Tenth Five Year Plan and Annual Plan 2002-2003*, Vol II. *op. cit.* p. 1381.
43. *Annual Plan 2004-2005, op. cit.* pp. 54-55.
44. Information about a *planted reed-bed* sewerage filter system is available from *Tency csr@auroville.org*. Details of Ecological sanitation are available at *http://Ecosan.org. members.freemail.absa.co.za/ernstt/*

DJB's primary object of management, meant that many of MCD's public toilets were often not getting their routine supply of water. "The problem MCD faces in cleaning toilets is due to the DJB cutting off water supply to some urinals."[45] Unclean toilets are a major factor behind defecation in public spaces near the toilets.

Even the unique initiative of this period, the construction of 956 CTCs under the Yamuna Action Plan, failed to rise above the level of a routine bureaucratic action. The original plan was for the construction of 1146 CTCs, but the MCD had difficulty in identifying the sites. Out of the sites selected, about 250 identified sites were on disputed land and 70 sites belonged to other government departments from whom clearance was required. Ultimately, construction could be taken up only on 956 sites. Even out of these 956 sites, 450 were sites where toilet complexes already existed. The existing 450 toilet complexes were demolished to make way for the construction of modern CTCs.[46] MCD officials had no hesitation in admitting that 450 toilet complexes were indeed being built where the old ones stood but claimed that the old toilet complexes had been reduced to rubble and garbage dump sites and what was more these were possibly among the best possible places in as much as they were the places to which the least objections were raised.

In many cases, MCD was unable to build toilets where its surveys showed a possibility of optimal use due to disputes about identity of the landowner whose permission was mandatory before construction could begin. There were 250 sites like the one near a graveyard in Sabholi in Shadhra North to which both the DDA and the Wakf Board laid a claim. In some cases where permission was not a problem, public resistance to having a toilet in the vicinity meant that construction could not take place. "No one wants a toilet block in front of their house or a park." Delay in getting permission from the appropriate authority in at least 70 cases reconciled the MCD to undertaking construction in available 'corners spaces or wherever practically feasible' where aesthetic and substantial use considerations became secondary. Since the project was funded by Japan, under a

45. "Poor Habits Raising a Stink, says MCD," (2001) *The Times of India*, October 27, Delhi Edition.

46. Pandey, Maneesh (2001). "MCD's 1,100 New Toilets include 450 Old Ones," *The Times of India*, December 26, Delhi Edition.

time bound project to be completed by 31st March 2002, the thrust of the project degenerated into "meeting the deadline and achieving the target number."[47]

Even if almost half the CTCs were constructed after demolishing old ones and many of the CTCs were not built at the best of locations, this endeavour was very significant considering that the 956 CTCs made available about 28,000 WC seats at a cost of about Rs. 140 crore within a period of about two years. The contribution of this venture becomes evident if one realises that in the Ninth Five Year Plan, only Rs. 662.50 lakh was provided under the Plan scheme of provision of Jan Suvidha Complexes and during the Plan period, 292 Jan Suvidha Complexes containing 7730 WC seats and 72 Mobile Toilet Vans containing 1008 WC were added by augmenting these funds from the Centrally Assisted Slum Development Scheme. For the Tenth Five Year Plan, Rs. 500 lakh only has been provided for this scheme.[48]

The four resettlement colonies where the mini sewage treatment plants had been installed had the potential of showcasing a substantial human waste management initiative that could have done Delhi proud. But this did not happen due to factors beyond the control of the MCD. In these resettlement colonies while the basic facilities were to be provided by the Slum and JJ Wing of the MCD, other facilities such as electricity, streetlights, transport, fair-price shops, schools and health centres were supposed to have been provided by the appropriate authorities before people moved in. Each of the families was legally entitled to plots between 12.5 and 18 square metres. The wastewater from the STP was a free resource available for upkeep of the public parks and private kitchen gardens. With such a large number of people living in close proximity in each part of the resettlement colony, each CTC could have evolved on the lines of a multipurpose toilet complex, pioneered by the Sulabh International where the toilet centred area becomes the venue for communication services, health care and even primary education.[49]

But as it happened the infrastructure was not in place when the squatter families were shifted. Many of the polluting factories that

47. *Ibid.*

48. *Tenth Five Year Plan of Delhi*, Urban Development, *op. cit.*, p. 1445.

49. Sulabh Public Toilets and Bath Complexes, Sulabh International Social Organisation.

were meant to be relocated in the same general areas due to the Supreme Court orders and which would have provided jobs did not do so. They held back due to the lack of infrastructure and the Delhi government's ambivalence in carrying out the Supreme Court's mandate due to its humanitarian concern for the "livelihoods of those engaged in these units."[50] The end result was that only a relatively few families were shifted from areas where a political patron with adequate clout to stall the move could not be mobilised. Many who came from old established clusters in areas such as Gole Market were accustomed to a regular income and basic infrastructure including possibly (in many cases) a free sewer based toilet. They were simply not up to complying with a pay and use toilet regime distracted as they were due to the disruption in their lives which the enforced relocation had created.

Because of the lower number of users of the CTCs than anticipated, the NGO contractors responsible for operating the CTCs were faced with a longer gestation period on their investment than the one they had bargained for. To reduce costs and save on the electricity bills, in which the charge was made on the enhanced rates for commercial enterprises, they kept the CTCs open during daytime only and not round the clock as agreed upon. This reduced the duration and scope of public access as per the contract with the MCD. There was reluctance on the part of the MCD to take any disciplinary measures against the NGO contractors because it had been unable to maintain its own side of the commitment of obtaining permission from the Central Ground Water Board for the contractors to draw on the ground water for use in the CTCs by drilling a bore well. This forced the NGOs to purchase water from private sources or hire water tankers, increasing the cost of their operations further. The users also considered the rates fixed for using the CTCs too costly. The MCD was aware of this feeling from the feed back it got from its monitoring unit, an integral part of the YAP project, and was willing to consider concessional family passes, but not a total waiver.[51]

Thus, a combination of reasons relating to 'issues such as lock of consensus on maintaining public health 'policies' such as total

---

50. *Tenth Five Year Plan of Delhi*, Urban Development, p. 1457.

51. Interview with Mahajan, S.K., Officer on Special Duty, Yamuna Action Plan, Phase II.

disinterest in reviewing an existing however paradigm and 'structures' such as the fact that the municipality is only one of the owners of public land, rather any single factor such as multiplicity of authorities turned a promising initiative to explore a more hygienic environmentally sound human waste management method into a lack lustre non-starter. Ideally funds for building 'public awareness' about environmental constraints made available under the Yamuna Action Plan should have been used to explore and create pilot project to verify claims made on behalf of on-site sewerless alternative sanitation technologies. Sulabh International's claims about using duck weed to clean waste water could have been investigated as also the efficacy of the Sulabh Themophilic Aerobic Composting (STAC) by which in 8-10 days biodegradable waste can become pathogen free manure without requiring physical handling.[52]

The Yamuna bank which was home to 60,000 squatters till recently, was the ideal site for demonstrating and testing, Kerala-based ecological sanitation expert, Paul Calvert's, ecosan toilets. These toilets were developed to meet the needs of people in coastal areas in Kerala, where the water table is high, possibility of flooding is continuous, pit latrines are impractical and sewers are out of question. Calvert's ecosan toilet is based on the principle of human waste segregation and recycling and 'closing the nutrient loop.' It consists of a slab built over two vaults. The slab has a hole over each vault for the faeces to drop in and a funnel-like device to collect the urine. The urine which has 90 per cent of waste nutrient is used directly as fertiliser. Before use, each hole is covered with straw to facilitate decomposition. The two holes are used on a rotational basis for six months. After six months, the decomposed faeces, is used as a soil conditioner. Calvert's system has three benefits. Pathogen laden excreta is prevented from causing infection, or polluting groundwater and it becomes a valuable resource for nourishing the soil. In the Kyoto 2003 Summit on Water and Sanitation, the potential of this sanitation system to provide a sustainable means for meeting millennium goal of providing 2.1 billion sanitation facilities by 2025 was recognised.[53]

---

52. Pathak, Bindeshwar (2003). "WSSD goals for water and sanitation: Making it happen through the Sulabh way." Paper presented at the *Delhi Sustainable Development Summit 2003*, February 6-9.

53. *www.ecosan.org*

Thus, from the point of view of sustainable solid waste management paradigm which in the context of human waste calls for hygienic, culturally appropriate ecologically sound practices, the initiatives of the Delhi government, DJB and MCD during 1998-2003 left much to be desired. However, there was some hope for the future, due to an extended opportunity for introspection by the concerned authorities. The Delhi government towards the end of this period initiated steps to transform the Delhi Jal Board into "an effective commercially oriented water utility capable of sustaining and improving the delivery of water supply and sewerage services" with the assistance from a World Bank Project. Consultants were appointed under the project to take up a comprehensive study of the department and prepare 'a feasible programme of immediate actions, works, supply of equipment, and technical assistance to be implemented in a period of four years.'[54] In a similar vein, with a view to adopt a more integrated approach for the development of slums, the Delhi government decided to study and prepare a Perspective Plan for taking care of the existing slum-dwellers and to prevent the growth of slums in the future.[55]

---

54. *4 Years, op. cit.*, p. 11.

55. *Tenth Five Year Plan of Delhi*, Urban Development, *op. cit.*, p. 1457.

6

Conclusion

Delhi has a solid waste profile which accurately reflects the transformation it has undergone since it became modern India's capital city. From a largely rural area in the early-1950s whose unique claim to a distinct status was that historically rulers had found it convenient to function from the same general area, it had, by the end of the century, become a sprawling, highly productive, ethnically diverse mega city with some of the most densely populated areas in the world within its urban limits. Today, while a substantial number of its citizens work in government agencies or in industry such as those relating to electrical goods, the large majority, especially those who migrated to Delhi in the 1980s and the 1990s, derive their livelihood from the tertiary sectors, from activities such as transport, construction and catering.

Delhi's households, institutions and centres of commerce and trade generate about 6000-7000 tonnes of waste in a day. Though paper and plastic due to their increasing use in packaging attract considerable attention, they constitute only 12 per cent of solid waste. 35 per cent consists of grit and 38 per cent of biodegradable or mainly food waste. Such a waste profile which is broadly in line with those of low income countries suggests that most consumers of goods and services in Delhi are relatively poor. Urban waste in India is conceptualised as a non-hazardous category. This is the case despite ample evidence that it is, in fact, quite hazardous. Healthcare centres and pathological laboratories co-dispose soiled bandages and used solvents etc. with their general waste. Chemical waste from factories is simply dumped on vacant land if not washed down into drains and sewers maintained by sub-city and city level authorities. Further, inadequate sanitation arrangements, not merely due to the underdeveloped infrastructure of government authorities but also due to technical and legal factors that inhibit action by them has meant that raw excreta is often got rid off by being placed inside and near public receptacles.

Ninety-four per cent of Delhi falls within the jurisdiction of the MCD and it is this body which is responsible for the regular collection, transportation and safe disposal of solid waste abandoned within its limits. Since the mid-1990s, with the publication of the report of the Nagpur based Environment Engineering Research Institute on the subject of improvements in the functioning of the MCD, official thinking in Delhi has been sensitised to the complexity of this task which requires managerial, financial institutional and attitudinal changes together with choices of appropriate rather than advanced technology and improvement in coordination with other government agencies. These suggestions were broadly in line with what had been indicated in Agenda 21, the blue print for operationalising the concept of 'sustainable development' which explicitly recognised that the management of waste, an unavoidable out come of development, was especially difficult problem that confronted the developing countries. Its heterogeneous character which posed a formidable challenge to those responsible for its management had no parallel in the developed world. This insight into its waste was indicated in the new paradigm on development which was endorsed by the world community in Rio de Janerio in the United Nations Conference on Environment and Development in 1992. According to this perspective all human beings in the present and the future 'were entitled to live a productive and healthy life in harmony with nature.' Urban planners in India have had no difficulty in recommending the application of the exhaustive suggestions relating to waste management in the Earth Summit action plan and treating these as a virtual foot-note to Article 48A of the Directive Principles of the Indian Constitution which calls on state agencies to 'protect and improve the environment' even as the other Principles in Part IV require that welfare policies be pursued side by side with those of growth.

A commitment at the national level towards SSWM where waste was understood as a broad category, and different government agencies cooperated with one another and with communities for regulating it while opting for a mix of sophisticated and low level technology for its scientific disposal was clearly evident in March 1999 in the perspective and the recommendations of the report of the Barman Committee. This committee had been appointed by the

Supreme Court to suggest improvements in the management of municipal solid waste in 300 Class I cities or urban centres that had more than one lakh residents. According to the Committee, public health could neither be expected to be preserved nor environment degradation halted unless steps were taken to regulate hazardous waste as well. Those in charge of health care facilities needed to apply themselves to the task of safe bio-medical waste disposal. Pollution Control Boards could ensure that designated sites were used for scientific disposal of industrial waste. More toilets had to be constructed to prevent open defecation which was a major cause of disease and environment pollution. The Committee was of the view that vigorous public awareness campaigns using information, education and communication were required to promote waste segregation reuse and recycling and composting activites. Suggestions for institutional innovations at the level of local government included involvement of ragpickers—hitherto regarded as thieves and obstructionists and holding of inter city meets for sharing experiences. A Technology Mission should be established by the Government of India under the Ministry of Urban Development, the Committee said, for a period of five years with the mandate to monitor the performance of various local bodies and to guide them about technology choices.

At the end of 1998-2003, in Delhi, the end-result of endeavours of central, state and local governments, was a considerable advance in the sustainable management of waste, though the progress was uneven across the four types of waste. It was greater in the case of bio-medical and non-hazardous waste as compared to hazardous and human waste. However, at the outset it needs to be recognised that all political parties that have governed at the Centre and in Delhi in the last decade seem to have been inspired by the notion of making Delhi both 'clean and green' by supporting the environment related initiatives of the Supreme Court. Delhi Government was able to benefit by the improvements that had occurred before it came into power in 1998. This trend was most clearly evident in the manner in which Delhi government was able to seem as a pioneer in the promotion of non-burn technology for biomedical waste in early 1999. A final decision on tenders for purchase of the autoclaves, which marked a radical break towards support of the discredited incinerators had been taken a few months earlier before it assumed power during the tenure of the previous BJP government.

A review of the circumstances in which the relevant steps, *viz.* the policy initiatives, pilot projects, and institutional reform related measures were taken show that though there was general improvement in solid waste management as compared to earlier times there was a differential in progress across the sectors which was due to at least two reasons. The disparity in progress in various areas of solid waste management is due to the fact that more non-government organisations were and are involved in pilot projects, research and advocacy in the first two areas as compared to the other two. Secondly, the national policy frameworks in relation to human and hazardous waste as compared to municipal and biomedical waste are weak and fragmentary.

Since January 2004, the increase in the number of twin green and blue bins in Delhi's streets and *'dhalaos'* for the collection of waste in segregated form suggests that MCD and Delhi government are fully committed about transforming the waste management scene in Delhi. Against this background, press reports that the bins are being misused and vandalized— there is 'mixed' bio-degradable and non-bio-degradable waste in both bins and bins are being repeatedly broken–should not be treated lightly. MCD needs to promote community level composting and waste management so that its advanced technology based initiatives at waste reduction do not destroy local livelihoods and cause neglect of cheaper tried and tested methods which are of a lower technological level but, which are also low cost and low energy consuming and capable of serving the desired goals of waste reduction and final scientific disposal. Since the Master Plan of Delhi for 2000-2021 is still to be finalised, a proposal by the NGO, Chintan, which calls for secure sites for ragpickers and waste merchant for sorting and storing waste deserves attention[1] if there has to be appropriate and adequate waste related infrastructure in future.

Given the well documented dangers to health due to plastic incineration official support to industries manufacturing plastic goods, such as by allowing import of poly vinyl chlorides (PVC) scrap should be conditional on these companies assuming responsibility for recycling plastic waste in accordance with the well known principle of Extended Producer Responsibility (EPR) until a cheap biodegradable

1. Chintan Environmental Research and Action Group (2004) *Space for Waste: Planning for the Informal Recycling Sector*, New Delhi.

plastic is synthesised and made readily available. Steps should be taken to implement the recommendations of the Barman Committee which noted that the recycling industry can be promoted through incentives such as priority allotment of land, power and water, tax holiday for the first few years besides preferential purchase of recycled products by government and semi-government bodies. In 1993 the Clinton administration issued an executive order requiring that all paper purchased for government agencies should contain 20 per cent or more post-consumer waste by 1995 increasing it to 25 per cent by 2000. The recycled paper industry received a huge boost since the US government was the world's largest paper buyer.[2]

The Barman Committee had noted that one of the most neglected aspects of MSW was the construction and demolition waste, left over bricks masonry pieces which in loose form or in a bag, are abandoned in public places. In Delhi, private contractors are frequently hired to carry away the waste supposedly to distant low lying areas not necessarily within the city limits. In reality much of it is abandoned in public spaces, or thrust into manholes. In the recent *Bhagidari* workshop *safai karamcharis* counted *'malba'* as an obstacle in the proper performance of their duties. MCD could by decentralising operations to the ward level consider following the example of the Chandigarh municipality which has been able to secure the cooperation of the city residents for its scheme of prompt collection by municipality staff of construction material in return for a service charge.[3]

Medical establishments in Delhi pose a greater threat to public health, perhaps by their very existence, rather than by their waste management policies since large numbers are unrecoganised. A leaf could be taken from the practice in the United States in drawing up the requirements for their legal existence. Ravi Agarwal has noted in his comparative study on biomedical waste in developing and developed countries, that as per the rules of the Environment Protection Agency, a hospital cannot be accredited unless it has a specific plan for managing hazardous materials and ensuring worker safety. Excise duties should be lowered on specific models of

2. Brown Lester R., (2001) *Eco-Economy, –Building an economy of the Earth*, Orient Longman, Hyderabad, p. 140.
3. Down to Earth (January 31, 2004) "Getting Wise about Waste—Chandigarh Shows the Way to Manage a City's Debris", Down to Earth, New Delhi, p. 44.

autoclaves so that the cost is comparable to small incinerators and the more benign disposal alternative gets preference by those starting private medical practice. Similar incentive should be there for digital devices so that unregulated exposure to mercury can be minimised.

In Delhi there has been a spurt in the output of hazardous waste in recent years because the city has become a major hub for dismantling operations of imported electronic devices. While some components such as IC chips are set aside for reuse, others may be burnt or soaked in acid to recover metals like copper, platinum and gold. The latter create health hazards for those who are engaged in these activities as also for those who live in surrounding areas which further strengthens the case for bringing recycling activities within the framework of law. United Nation studies have shown that manufactured goods require raw materials which is many times their weight for construction purposes—in the case of computers it is ten times and in the case of a car it is two times. Since there is a saving in both raw materials and reduction of waste when feed stock is derived from waste materials by industry, a wide variety of waste both hazardous and non-hazardous is legally imported in India. Hazardous wastes can be imported by registered handlers and reprocessors subject to conditions imposed by the Central Pollution Control Board. The Directorate General of Foreign Trade (DGFT) under the Union Ministry of Commerce and Industry is authorised to issue licenses for importing non-hazardous waste. In recent years, there has been a rise in instances of unauthorised entry of hazardous waste loads entering the country. Further, the quantities of non-hazardous waste that is imported and its bewildering variety ('from steel to incineration ash to pig hair') from as many as 105 countries, including the poorest, has grown sharply.[4] As the High Power Committee appointed by Supreme Court noted, the MoEF as the overarching authority responsible for enforcing the Hazardous Waste Rules has simply not been able to put in place structures whether it was in terms of rules, personnel or technology to regulate the form in which the waste was imported or the method of its utilisation. Harmful impacts could have been mitigated if there was a secure landfill for the final residue. There was none in Delhi until December 2003. A suitable site could not be agreed upon by concerned authorities.

4. Down to Earth (2004). "Just Dump It", Down to Earth, Nov. 15, 2004, p. 30.

By the end of June 2004, in Delhi, 945 cases of cholera were reported as compared to 375 and 485 at the end of 2000 and 2001 respectively, and experts warned that the disease could break out in epidemic proportions unless the DJB prevented the contamination of the water supply.[5] The DJB was understandably on the defensive because a major source of contamination of the water was make shift sewers since the repair work on the main lines was lagging behind schedule.[6] The urgency of expediting research for an economical on-site 'green' toilet for urban conditions has never been more apparent than it is now. In seminal form one such device already exists in the 'ecosan' ecological sanitation toilet invented by Paul Calvert. The Kerala based sanitation expert has exhorted governments and corporates for assistance in modifying and refining the toilet that has since 1995 provided triple benefits of hygienic sanitation, prevention of contamination of water bodies and return of nutrients to the soil, in rural areas of southern India and in Sri Lanka.[7] Similar assistance should be made available to NGOs and academic institutions engaged in pilot projects on waste water treatment using solar energy and biocultures such as EM (Effective Microbes), which consists of a combination of bacteria and enzymes.[8]

In recent years, a consequence of globalisation has been the trend towards conscious sharing of views among larger cities in different geographic regions on the ground that they have more in common with one another than with smaller urban centres in their own countries. In the context of advance towards a sustainable waste management model in the shortest possible time period, focus should be on investigating waste related technologies which have been adapted to conditions such as high population density low energy and limited public fund availability that commonly exist in developing countries. An example is the 'vacutug,' a 200 litre sewage suction device used in Dhaka, a city of over 10 million people, which has sewers only in one third of its total area. The national organisation Dushtha Shasthya Kendra (DSK) working in collaboration with an international NGO Water Aid uses this device which looks like a bath

5. Hindustan Times (2004). "Sarita Vihar Tests Positive for Cholera", *Hindustan Times*, July 1.
6. Hindustan Times (2004). "Sewage Flows Back in West Delhi Homes" *Hindustan Times*, June 26.
7. Calvert, Pual (February 29, 2004) "Homegrown Remedy", *Down to Earth*, p. 3.
8. Joshi, Sandeep (2004) (February 15, 2004). "River Cleaning Costs a Lot," *Down to Earth*, p. 2. EM is marketed by Organics Solution, N-38 Panchsheel Park, New Delhi.

tub on wheels and functions as a satellite of the main system which consists of a 1900-litre-capacity tank. The former can reach areas with very narrow access and can even be maneuvered over obstacles. It pumps the waste back through a 75mm diameter hose to the main tank. While the main unit can handle up to 1900 litres of sewage and is best for the larger toilet complexes, the satellite unit can be moved through narrow streets to reach the smallest toilet complex. Within minutes it can suck out the load, which would have taken a person many hours.[9] Such an innovation has special significance in India since the Government of India is committed through an Act of Parliament in 1993 to end the demeaning and hazardous practice of scavenging which affects more women than men. The practice exists because there are no affordable mechanical substitutes.

Adoption of Charters which explicitly spell out the work ethic to which employees are committed by as many as 66 Departments, and the passage of the Right to Information Act 2000 with special citizen friendly features are recent measures which the government of Delhi has instituted. The object is to make the structures of governance more responsible and responsive, a primary consideration of the sustainable waste management model. Against this background, media reports that Delhi government is contemplating an amendment to reduce the scope on which questions can be asked are a matter of concern. With a Congress party led coalition in power at the Centre, a bill to grant statehood to Delhi in which the lieutenant governor has fewer powers than at present, and the Delhi government has powers over all areas except the VIP zone under the NDMC is considered likely. This will facilitate a holistic approach to the management of Delhi's waste in the coming years especially if the Master Plan for Delhi 2000-2021 as suggested by the National Planning Board for Delhi Region can provide a detailed plan on how the development of surrounding towns can be synchronised with that of Delhi.

However, if there is to be steady visible progress on the ground in the implementation of the MSW Management and Handling Rules as distinct from the minimal advance required to prevent show cause notices from the Supreme Court for neglecting implementation of the law, then the CSE division of the MCD should decentralise collection and transportation arrangements to the ward level by increasing the

9. Return of the Vacutugs-Bangladesh.tve.org

number of Assistant Executive Engineers, Sanitary Inspectors, and Sanitary Guides. The senior AE in charge of the ward should call for registration of self designated RWAs, market associations, professional and educational institutions medical establishments, NGOs and waste dealer associations within the ward and hold periodic meetings with their representatives chosen through the drawing of lots on how waste can be prevented/reduced and disposed of locally so that only the minimal amount is dispatched to the *dhalaos*. The councillor who represents the ward in the MCD should be regarded as an ex-officio member of this group. The MLA and the MP within whose area the ward exists should be invited to be members. MCD should draw upon its property tax collection to facilitate zero garbage zones and announce discounts in property tax levy to collectivities in return for reduced garbage generation. The AE should seek out and enroll potential members of the Ward Solid Waste Management Association in the manner in which Delhi Police relates to householders in the context of its periodic servant verification drives. The agenda, time and venue of the meetings should be circulated well in advance and the resolutions and progress reports made public. At least two types of obstructions on Delhi roads should get reduced with this locality centered waste management approach. There will be fewer stray cattle and dung heaps when the importance of fresh cow dung for decomposing horticulture waste on site becomes better known. There will be fewer mounds of rubble if one man's construction and demolition waste becomes another's resource for leveling a low lying area. This initiative by the MCD can pave the way for a review of the existing 12 large, unwieldy and dysfunctional Ward Committees in Delhi created under the 74th amendment. In their present form these bodies have failed to meet expectations. They were meant to be authentic grass-root level structures which are required to include "persons having special knowledge or experience in Municipal administration" and by constitutional law they have the authority to identify priorities of the ward and to plan and supervise municipal works. Since the MCD is a vast organisation, the boundaries of some of the 134 wards should be redrawn so that the population is more evenly divided and there can eventually and ideally be a separate Committee for every Ward as is the case in Thiruvananthapuram and Kolkata.[10]

10. Singh B.N. and Shipra Maitra (2000). "Formation of Ward Committees" in *Urban Governance in India*, pp. 87-96. The suggestions in this section have been adapted from their formulations.

Delhi is a microcosm of India. As its first city it is in a position of natural leadership in demonstrating how urban centres can improve the quality of life of their ordinary citizens even as they cope with the daunting challenges of maintaining the momentum of economic growth in the present era of globalisation through new processes and use of substances whose properties are not fully understood. The custodians of Delhi, Government of India, Delhi government and the sub Delhi authorities, particularly the MCD and the DJB should move in tandem to consolidate the beginnings that have been made in recent years to manage solid waste in a manner that is environment friendly, inclusive of all stakeholders, supportive of both the precautionary and 'polluter must pay' principles and of appropriate technologies upon which local livelihoods depend.

Bibliography

Acts and Rules

The Constitution of India, Fourth Edition, Universal Law Publishing, Delhi.

Delhi Plastic Bag (Manufacture, Sales and Usage), and Non-Biodegradable Garbage (Control) Act, 2000.

Delhi Municipal (Amendment) Act, 1993, Biomedical Waste (Management and Handling) Rules, 1998, *www.ceeraindia.org*

Biomedical Waste (Management and Handling), (Second Amendment), Rules, 2000, *www.ceeraindia.org*

Hazardous Waste (Management and Handling), Rules, 1989, *www.ceeraindia.org*

Hazardous Waste (Management and Handling), Amendment Rules, 2000, *www.ceeraindia.org*

Hazardous Waste (Management and Handling), Draft Rules, 2002, *www.ceeraindia.org*

Municipal Solid Waste (Management and Handling), Rules, 2002, *www.ceeraindia.org*

Recycled Plastics Manufacture and Usage Rules, 1999, *www.ceeraindia.org*

Official Reports/Documents

Agenda 21 (1992). United Nations Department of Economic and Social Affairs, Division for Sustainable Development, *www.un.org/esa/sustdev*

Agenda 21: An Assessment (2002). Ministry of Environment and Forests, Government of India, *www.envfor.nic.in*

Annual Plan of India (1999-2000). *www.planningcommission.nic.in*

Appraisal of Ninth Five Year Plan (1997-2002) and Annual Plan 2001-02, Planning Department, Government of Delhi. *www.delhiplanning.nic.in*

Asia-Pacific Environment Outlook 2 (2001). UNEP, Thailand.

Audit Report on NCT of Delhi (2002). *www.cag.nic.in*

———. (2003). *www.cag.nic.in*

Census of India (2001). *www.censusindia.net*

DDA (Delhi Development Authority) (1962). *Master Plan*, New Delhi, 1962, DDA, New Delhi.

———. (1990). *Master Plan for Delhi*, Perspective 2001, DDA, New Delhi.

Delhi 21, Delhi Urban Environment and Infrastructure Improvement Project (DUEIIP) Report (January 2001). Government of India, Ministry of Environment and Forests and Government of National Capital Territory of Delhi, Planning Department.

Economic Survey of Delhi (1999-2000). *www.delhiplanning.net.in*

———. (2001-2002). *www.delhiplanning.net.in*

Fourth Economic Census of Delhi (1998). Directorate of Economics and Statistics, Government of National Capital Territory of Delhi, New Delhi.

Fifth Srishti Survey of Medical Waste Disposal Practices in Health-care Units of Delhi (2000). Srishti.

High Power Committee Report on Management of Hazardous Waste (2001). *www.cpcb.delhi.nic.in/hpc report*

Municipal Corporation of Delhi, *Background Note on Conservancy and Sanitation Engineering Department.*

National Environment Engineering Research Institute (NEERI) (August 1995). *Strategy Paper on SWM in India*, Nagpur.

———. (1996). *Solid Waste Management in MCD Area, Phase 1 Report, Action Plan for Immediate Improvement*, Nagpur.

Report of the High Powered Committee (1995). *Urban Solid Waste Management in India*, Planning Commission, Government of India, New Delhi.

Report of the Committee constituted by the Hon. Supreme Court of India (1999). *Solid Waste Management in Class 1 Cities in India.*

Report on Recycling Responsibility, Traditional Systems and New Challenges of Urban Waste Management in India, (June 2002), Srishti, New Delhi

Report on the Census of Industrial Units in Approved Industrial Areas of Delhi, (2002), National Council of Applied Economic Research, Delhi, *www.ncaer.org*

Rio Declaration on Environment and Development (1992). United Nations Department of Economic and Social Affairs, Division for Sustainable Development, www. un.org/ esa/sustdev.

State of Environment Report for India (2001). *www.envfor.nic.in*

State of Environment Report for Delhi 2001 (2001). The Energy Research Institute, New Delhi.

Sulabh Public Toilets and Bath complexes, Sulabh International Social Organisation.

The Energy Research Institute, Final Report (draft) No. 1999EE41 (March 2002). Urban Services Environmental Rating System.

Tenth Five Year Plan of Delhi, Planning Department, Government of Delhi, *www.delhiplanning. nic.in*

Twenty Fifth Bhagidari Workshop on Municipal Solid Waste Segregation (18-19-20th December 2003).

4 Years (2002). Working Report, Delhi Government.

White Paper on Pollution in Delhi with an Action Plan (1997). *www.envfor.nic.in*

Working Report on Bhagidari, *The Citizen-Government Partnership* (2001). Government of India

Yamuna Action Plan, *www.yap.nic.in*

Books and Papers

Agarwal, Ravi (1998). *Medical Waste, Issues, Practices and Policies*, Srishti, New Delhi.

Agarwal, Ravi and Sanjay K. Gupta, *Rethinking Urban Waste in India: Creating a Community Paradigm*, Social Change.

Ali, Mansoor, Andrew Cotton, and Ken Westlake (1999). *Down to Earth*, Loughborough University, UK.

Beaton, Russ and Chris Maser (1999). *Reuniting Economy and Ecology in Sustainable Development*, Lewis Publishers, London.

Bell, Simon and Morse, Stephen (2003). *Measuring Sustainability—Learning from Doing*, Earthscan Publications Ltd., London

Brockerhoff, Martin P. (2000). "An Urbanising World", *Population Bulletin*, Vol. 55, No.3, *www.prb.org*

Bisson, Katy and John Proops (ed.)(2002). *Waste in Ecological Economics*, Edward Elgar Publishing, UK.

Biswas, Dilip (1997). *Environment Legislation: Challenges of Enforcement*, www.cleantechindia.com

Beukering, Pieter Van, M. Sehker, R. Gerlagh, Vijay Kumar (1999). "Analysing Urban Solid Waste in Developing Countries: A Perspective on Bangalore, India," *CREED Working Paper No. 24*, Amsterdem, *www.waste.nl*

Booth, Beverley *et al.*, (2001). *Urban Health and Development—A Practical Manual for Use in Developing Countries*, Macmillan, London.

Bruntland, G. (ed.) (1987). *Our Common Future: World Commission on Environment and Development*, Oxford University Press, Oxford.

Chary, S.N. and Vinod Vyaslu (ed.) *Environmental Management.*

Chaplin, Susan E. (1999). "Cities, Sewers and Poverty: India's Politics of Sanitation," in *Environment and Urbanisation*, Vol. 11, No. 1.

Chaturvedi, Bharati (1998). *Public Waste Private Enterprise*, The Heinrich Boel Stiftung, Berlin.

———. (September 2003). "Waste-handlers and Recycling in Urban India: Policy, Perception and the Law," *Social Change*, June: Vol.33 Nos. 2&3.

Chintan Environmental Research and Action Group (2004). *Space for Waste: Planning for the Informal Recycling Sector*, New Delhi.

Coad, Adrian, (ed.) (1995). *Lessons From India in Solid Waste Management*, Sponsored by the Department of International Development, Government of UK.

Cointreau-Levine, Sandra (1982). *Environmental Management of Urban Solid Wastes in Developing Countries*, A Project Guide for the World Bank staff.

———. (1994). *Private Sector Participation in Municipal Solid Waste Services in Developing Countries*, Volume 1, The Formal Sector, The World Bank, Washington.

Durand-Lasserve, A. and L. Royston (ed.) (2002). *Holding their Ground: Secure Land Tenure for the Urban Poor in Developing Countries*, Earthscan, London.

Dresner, Simon, (2002). *The Principles of Sustainability*, Earthscan Publications Ltd., London.

Ghosh, Archana (2000), *Solid Waste Management in Delhi, An Exploratory Study on Local Government-Community Interface*, Institute of Social Science, New Delhi.

Gulaty, Sachin and K.V. Santhosh Ragavan (26 April 2003). "Solid Waste Management in India," Background Paper prepared for Workshop *The Need for Institutionalisation of Solid Waste Management in India*, The Energy Research Institute, New Delhi.

Hardoy, Jorge E., Diana Mitlin, and David Satterthwaite (2001). *Environmental Problems in an Urbanising World*, Earthscan Publications Ltd, London and Sterling, VA.

Joshi, B.L. (2003). "Census of Industrial Units in Approved Industrial Areas of Delhi", *NCAER Industry Briefs 1, www.ncaer.org*

Jha, Nimesh, *Urban Growth and Financing of Municipal Services: A Case Study of Municipal Corporation of Delhi* 1985-99.

Kundu, Amitabh (16-17 June, 2003). "Population Dynamics and Economic Development in Delhi," *Seminar on Sustainable Development of Delhi*, New Delhi.

Liu, David H.F and Bela G. Liptak (ed.) (1996). *Hazardous Waste and Solid Waste*, Lewis Publishers, London.

Low, Nicholas, Elander, Ingemar, Gleeson, Brendan and Rolf, Lidskog (Ed.) (2000). *Consuming Cities: The Urban Environment in the Global Economy After the Rio Declaration*, Routledge, London.

Mani, Shyamala K., *et al.*, *Less Plastics for a Better Future*, Centre for Environment Education, Delhi.

———. (2002). *Combating Solid Waste—A Fruitful Strategy*, Centre for Environment Education, Delhi.

Mathur, Kuldeep (2003). "Battling For Clean Environment: Supreme Court, Technocrats and Populist Politics in Delhi," Seminar on *Sustainable Development of Delhi*, Institute for Human Development, New Delhi (June 17-18).

McDougall, Forbes *et al.* (2001). *Integrated Solid Waste Management—A Cycle Inventory*, Blackwell Sc Limited, Oxford.

Mehra, Ajay K. (2003) "Governing India's First City," Seminar on *Sustainable Development of Delhi*, Institute for Human Development, (June 16-17).

Orlando, Laura (2000). *Sustainable Sanitation, Part 3*, The Resource Institute for Low Entropy Systems, Boston, *www.riles.org*

Palnitkar, Sneha, (ed.), *Manual on Solid Waste Management*, All India Institute of Local Self Government, Mumbai.

Pathak, Bindeshwar (February 6-9, 2003) "WSSD Goals for Water and Sanitation: Making it Happen Through the Sulabh Way" Paper presented in the Delhi Sustainable Development Summit, at India Habitat Centre (February 6-9).

Pathak, Bindeshwar (June 16-17, 2003). "Sustainable Sanitation Technologies: Sulabh Experience", Seminar on *Sustainable Development of Delhi: Emerging Challenges and Policy Options in the Era of Globalisation*, Institute of Human Development, New Delhi (June 16-17, 2003).

Pinto, M. (2000). *Metropolitan City Government in India*, Sage Publications, New Delhi.

Plummer, Janelle (1999). *Municipalities and Community Participation*, Earthscan, London.

Richards, P.J. and A.M. Thomson (eds.) (1999). *Basic Needs and the Urban Poor: The Provision of Command Services*, Crown Hamilton.

Ruet, Joel, *et al.*, (2002) *Water and Sanitation Scenarios in Indian Metropolitan Cities-Resources and Management in Delhi, Calcutta, Chennai and Mumbai*, CSH Occasional Paper No 6, French Research Institute.

Satterthwaite, David (April 2001). "Reducing Urban Poverty: Constraints on the Effectiveness of Aid Agencies and Development Banks and Some Suggestions for Change," *Environment & Urbanisation*, Vol.13, No. 1.

Schubeler, P. *et al.*, (1996). "Conceptual Framework for Municipal Solid Waste Management in Low Income Countries," World Bank Urban Management Programme, *Working Paper No. 9*, *www.worldbank.org/publications/*

Sinha, A.H., Maqsood Md., and Iftekhar Enayetullah (ed.), *Community Based Solid Waste Management: The Asian Experience.*

Sulabh Public Toilets and Bath Complexes, Sulabh International Social Organisation.

Svadlenka, Robert (2002). "Sustainable Sanitation Approaches for the Urban Poor", *www.worldhungeryear.com*

Upadhayay, S. and V. Upadhayay (2002). *Environment Protection: Land and Energy Laws*, Vol.III, Lexis Nexis, New Delhi.

van de Klundert, Arnold and Anschutz, Justine (1999). "Integrated Sustainable Waste Management: The Selection of Appropriate Technologies and the Design of Sustainable Systems is not (only) a technical issue," CEDARE/IETC Inter-Regional Workshop on Technologies for Sustainable Waste Management, Alexandria, Egypt, *www.waste.nl*

———. (2002). *Integrated Sustainable Waste Management-the Concept.*

Venkateswaran, Sandhya (1995). *Alternative Solid Waste Management Systems, Eight Case Studies from Across the Country*, National Institute of Urban Affairs, New Delhi.

———. (1996). *Community Sanitation, Alternative Solid Waste Management Systems*, National Institute of Urban Affairs, Delhi.

Wasser, Otter (2002). "Sustainable Sanitation", *www.otterwasser.de*

Magazines, Newspapers, Journals

Aggarwal, Manu (2003). "City backyards." *Down to Earth*, (15 June 2003).

Anu Goel (2002). "Small price – Big returns," Srishti's factsheets on 14 priority areas, Srishti (June).

Agarwal, Ravi (2000). "Foreword," *Managing Hospital Waste*, Srishti (September).

———. (2002). "Slowly, slowly, down the stream", *Indian Express*, Delhi Edition (17 May).

Chaturvedi, Bharati (2000). "Plastic not so fantastic," *Hindustan Times*, Delhi Edition (9 September).

Down To Earth (2004). "Just dump it." (Nov. 15).

———. (2004) "CETP standoff." (April 30).

———. (2004) "Homegrown Remedy." (February, 29).

———. (2003). "Prickly Issue." (July 31).

———. (2003). "Black liquor". (June 15).

———. (2003). "Going to pot." (June 15).

———. (2003). "CETPs: In fashion." (15 January).

———. (2003). "For a good churn." (15 January).

———. (2003). "The pollution commons." (15 January).

———. (2002). "Sewer systems totally destroy nature's nutrient cycle." (February 28).

———. (2002). "Thinking clean." (February, 28).

Environment (2002). Delhi Government (November).

Jayaraman, Nitynand (2000). "Hazardous Wastes", *The Hindu Survey of the Environment* 2002.

Mago, Chandrika (2001). "DJB sewage treatment plants sloppily managed, says CPCB," *The Times of India*, Delhi Edition, Delhi (12 October).

Nadkarni, Manoj (2002). *Down to Earth*, "Drowning in Human Excreta." (28 February).

Pandey, Maneesh (2001). "MCD's 1,100 new toilets include 450 old ones," *The Times of India*, Delhi Edition (26 December).

Roy, Subhajit (2003). "10 years on, Yamuna still toxic," *Times News Network* (2 June).

Sehgal, Omesh (2003). "Without some more powers, Delhi may soon reach the point of no redemption," *Hindustan Times*, Delhi Edition (24 January).

Sharma, Aruna P. (2002). "Lack of coordination plagues Bhajanpura," *Hindustan Times*, Delhi Edition (4 April).

———. (2003). "HC order blocks shifting of slums," *Hindustan Times*, Delhi Edition (8 January).

Sharma, Vibha (2001). *Hindustan Times*, Delhi Edition (July 9).

Sridhar, Lalitha, "Lack of sanitation causes distress and death," *www.infochangeindia.org*

TERI Seminar, 2003.

TERI vision, (2001) Fundamental problem of waste generation and disposal in Delhi (September), *www.teriin.org/news/terivsn/issue*

The Times of India, Delhi Edition (27 October 2001). "Poor habits raising a stink, says MCD,"

———. (2001). "More financial powers for state Govt." (10 September).

———. (2001). "Campaigns fail to deliver." (26 December).

Times News Service (2002). "It's poison everywhere." (15 April).

Toxic Links. "Problem of Medical Waste." Toxic Links_files/issues

33tve, *Return of the Vucutugs-Bangladesh*, *www.tve.org*

Srishti, (2002). Hospital Waste, Srishti's factsheets on 14 priority areas (June).

Verma, R.K. (2002). "Straight Question," *Hindustan Times*, Delhi Edition (29 March).

Vohra, Pankaj (2002). "What will my detractors say next." (24 November).

Walia, A.K. (2000). Delhi's Health Minister as reported in *Hindustan Times*, Delhi Edition (8 February).

Index

C

D

E